Grade 5

T5-BSK-999

Discovery Education | SCIENCE TECHBOOK

Unit 1
What Is Matter Made Of?

Copyright © 2020 by Discovery Education, Inc. All rights reserved. No part of this work may be reproduced, distributed, or transmitted in any form or by any means, or stored in a retrieval or database system, without the prior written permission of Discovery Education, Inc.

NGSS is a registered trademark of Achieve. Neither Achieve nor the lead states and partners that developed the Next Generation Science Standards were involved in the production of this product, and do not endorse it.

To obtain permission(s) or for inquiries, submit a request to:

Discovery Education, Inc.
4350 Congress Street, Suite 700
Charlotte, NC 28209
800-323-9084
Education_Info@DiscoveryEd.com

ISBN 13: 978-1-68220-803-8

Printed in the United States of America.

5 6 7 8 9 10 CWM 26 25 24 23   B

Acknowledgments

Acknowledgment is given to photographers, artists, and agents for permission to feature their copyrighted material.

Cover and inside cover art: Stewart Makin / EyeEm / Getty Images

Table of Contents

Unit 1: What Is Matter Made Of?

Letter to the Parent/Guardian..................................vi

Unit Overview 1

Anchor Phenomenon: Water Evaporating from a Fishbowl..........2

Unit Project Preview: Decreasing Water Levels 4

Concept 1.1 Describing Matter in Words and Numbers

Concept Overview 6

Wonder..8

Investigative Phenomenon: Hands and Hot Chocolate..........10

Learn..14

Share..40

Concept 1.2 Changes to Matter

Concept Overview 48

Wonder...50

Investigative Phenomenon: Melting Matter....................52

Learn..58

Share..94

Concept 1.3 A Model of Matter

Concept Overview 102
    Wonder.. 104
        Investigative Phenomenon: Models 106
    Learn 110
    Share 126

Unit Wrap-Up

Unit Project: Decreasing Water Levels 136

Grade 5 Resources

Bubble Map R3
Safety in the Science Classroom R4
Vocabulary Flashcards R7
Glossary R23
Index R54

Unit 1: | What Is Matter Made Of?

Dear Parent/Guardian,

This year, your student will be using Science Techbook™, a comprehensive science program developed by the educators and designers at Discovery Education and written to the Next Generation Science Standards (NGSS). The NGSS expect students to act and think like scientists and engineers, to ask questions about the world around them, and to solve real-world problems through the application of critical thinking across the domains of science (Life Science, Earth and Space Science, Physical Science).

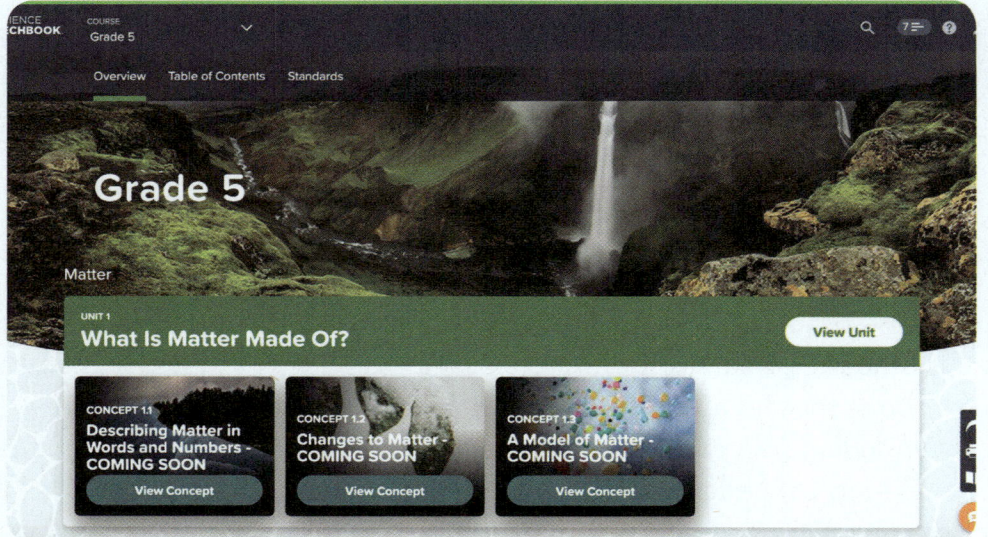

Science Techbook is an innovative program that helps your student master key scientific concepts. Students engage with interactive science materials to analyze and interpret data, think critically, solve problems, and make connections across science disciplines. Science Techbook includes dynamic content, videos, digital tools, Hands-On Activities and labs, and game-like activities that inspire and motivate scientific learning and curiosity.

You and your child can access the resource by signing in to www.discoveryeducation.com. You can view your child's progress in the course by selecting the Assignment button.

Science Techbook is divided into units, and each unit is divided into concepts. Each concept has three sections: Wonder, Learn, and Share.

Units and Concepts Students begin to consider the connections across fields of science to understand, analyze, and describe real-world phenomena.

Wonder Students activate their prior knowledge of a concept's essential ideas and begin making connections to a real-world phenomenon and the **Can You Explain?** question.

Learn Students dive deeper into how real-world science phenomenon works through critical reading of the Core Interactive Text. Students also build their learning through Hands-On Activities and interactives focused on the learning goals.

Share Students share their learning with their teacher and classmates using evidence they have gathered and analyzed during Learn. Students connect their learning with STEM careers and problem-solving skills.

Within this Student Edition, you'll find QR codes and quick codes that take you and your student to a corresponding section of Science Techbook online. To use the QR codes, you'll need to download a free QR reader. Readers are available for phones, tablets, laptops, desktops, and other devices. Most use the device's camera, but there are some that scan documents that are on your screen.

For resources in Science Techbook, you'll need to sign in with your student's username and password the first time you access a QR code. After that, you won't need to sign in again, unless you log out or remain inactive for too long.

We encourage you to support your student in using the print and online interactive materials in Science Techbook, on any device. Together, may you and your student enjoy a fantastic year of science!

Sincerely,

The Discovery Education Science Team

Unit 1: | What Is Matter Made Of?

Unit 1
What Is Matter Made Of?

Get Started

Water Evaporating from a Fishbowl

Aunt Jenna notices decreasing water levels in her fish tank. In this unit, you will describe and measure properties of common materials like water, glass, plants, and rocks. At the end of the unit, you will be able to develop a model of the three states of matter and use it to explain what happens when water evaportates.

Quick Code: us5006s

Water Evaporating from a Fishbowl

Think About It

Look at the photograph. **Think** about the following questions.

- How can we tell different materials apart?
- How do materials change when they dissolve, evaporate, melt, mix together, or are heated?
- How can we model the differences among solids, liquids, and gases?

Fishbowl

Unit 1: What Is Matter Made Of?

Unit Project Preview

 **Solve Problems Like a Scientist**

Unit Project: Decreasing Water Levels

In this project, you will use what you know about matter and its properties to investigate and explain decreasing water levels in a fishbowl over time.

Quick Code: us5007s

Fishbowl

| **SEP** | Planning and Carrying Out Investigations |
| **CCC** | Cause and Effect |

Ask Questions About the Problem

You are going to design and complete an investigation into how water evaporates from a fishbowl using what you know about matter and its properties. **Write** some questions you can ask to learn more about the problem. As you learn about matter in this unit, **write** down the answers to your questions.

Unit 1: What Is Matter Made Of?

CONCEPT 1.1

Describing Matter in Words and Numbers

Student Objectives

By the end of this lesson:

- [ ] I can develop models that show very small and very large quantities of particles in different states.
- [ ] I can classify materials based on their properties and describe patterns in the properties of similar materials.
- [ ] I can communicate how the structure of a material and its function are related.
- [ ] I can investigate the thermal conductivity of different materials.
- [ ] I can create and evaluate multiple designs of a hot beverage container that is comfortable to hold.

Key Vocabulary

- [ ] gas
- [ ] liquid
- [ ] mass
- [ ] material
- [ ] matter
- [ ] measure
- [ ] particle
- [ ] property
- [ ] scale
- [ ] solid
- [ ] state of matter
- [ ] substance
- [ ] volume

Quick Code: us5009s

Activity 1
Can You Explain?

How is matter described, measured, and classified?

_____
_____
_____
_____
_____
_____
_____
_____
_____
_____
_____

Quick Code: us5010s

1.1 | Wonder How is matter described, measured, and classified?

Activity 2
Ask Questions Like a Scientist

Quick Code: us5011s

Hands and Hot Chocolate

Look at the photograph. Then, **answer** the questions.

Let's Investigate Hands and Hot Chocolate

SEP Asking Questions and Defining Problems
CCC Cause and Effect

How does it feel to hold a mug of hot chocolate? Why does it feel that way? **Write** three questions you have, and **share** them with the class.

I wonder...

I wonder...

I wonder...

Concept 1.1: Describing Matter in Words and Numbers | 11

1.1 | Wonder
How is matter described, measured, and classified?

Activity 3

Evaluate Like a Scientist

What Do You Already Know About Describing Matter in Words and Numbers?

Quick Code: us5012s

Measuring Matter

Look at the pictures. Which tool would you use to measure volume? Which would you use for length? Which would you use for temperature?
Draw a line from each picture to the property measured.

Flask

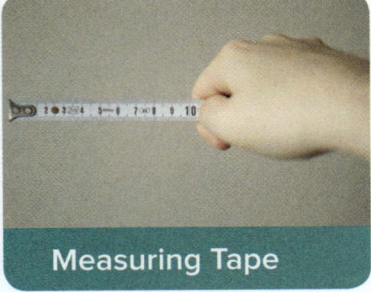
Measuring Tape

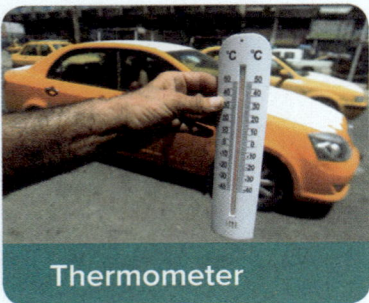
Thermometer

- Length
- Volume
- Temperature

What Is Matter?

Circle the word that best describes what all matter is made of.

forces   light   particles   solids

Discuss with Your Class

What tools can be used to measure the properties of materials?

Tools	Properties

Why is it useful to measure different properties?

_____
_____
_____
_____
_____

Concept 1.1: Describing Matter in Words and Numbers

1.1 | Learn How is matter described, measured, and classified?

What Is Matter?

 **Activity 4**
Observe Like a Scientist

What's the Matter?

Complete the Properties portion of the interactive. **Examine** the characteristics of each object. Then, **choose** two objects from the interactive, and describe their properties of size, shape, color, texture, and hardness.

Quick Code: us5013s

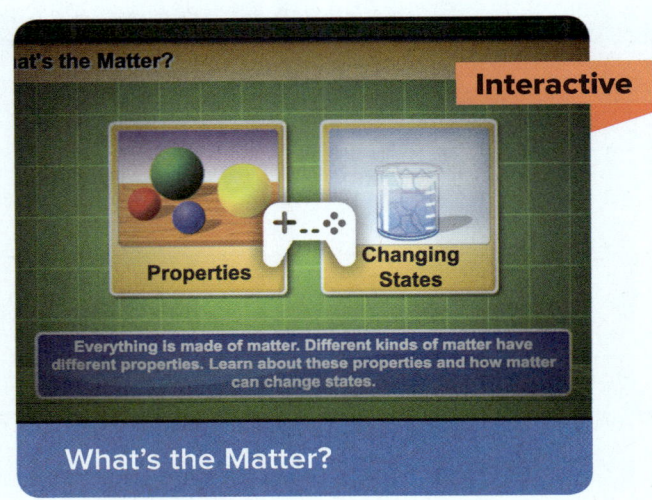
What's the Matter?

Objects

Wood Block	Video Game	Toy Truck
Baseball	Basketball	Tennis Ball

SEP Obtaining, Evaluating, and Communicating Information

Write the name of the object. Then, **mark** with an X the place on each scale that best describes the object. Don't worry about being exact.

Object 1 _____

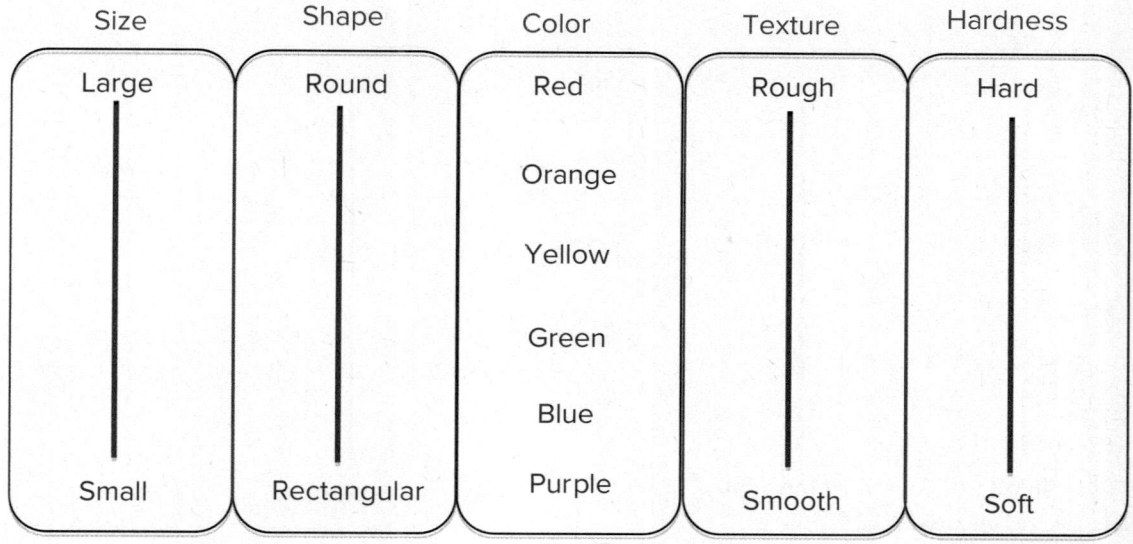

Concept 1.1: Describing Matter in Words and Numbers

1.1 | Learn How is matter described, measured, and classified?

Write the name of the object. Then, **mark** with an X the place on each scale that best describes the object. Don't worry about being exact.

Object 2 _____

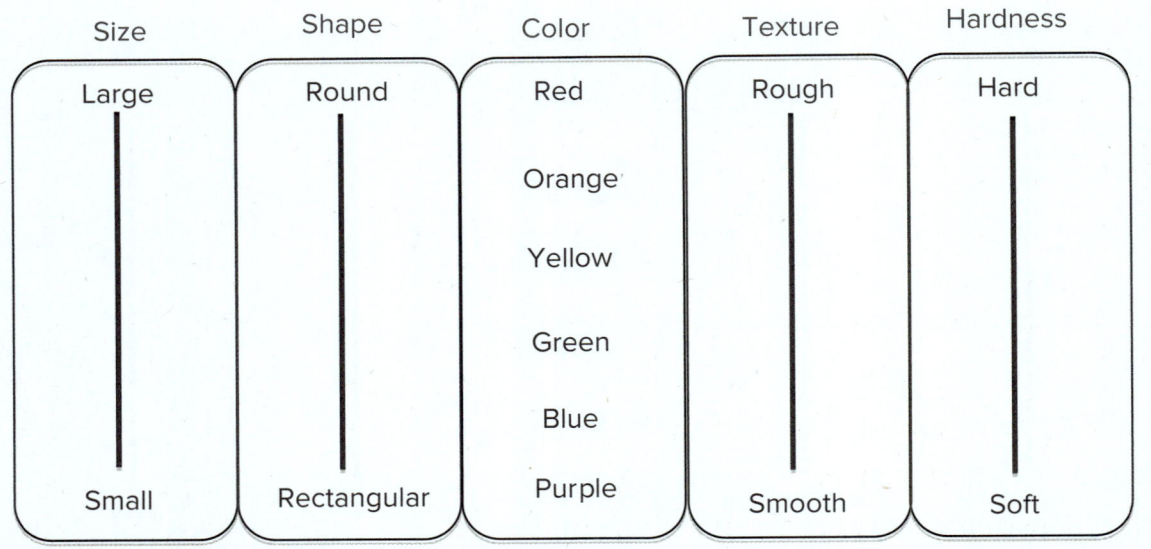

Other Object _____

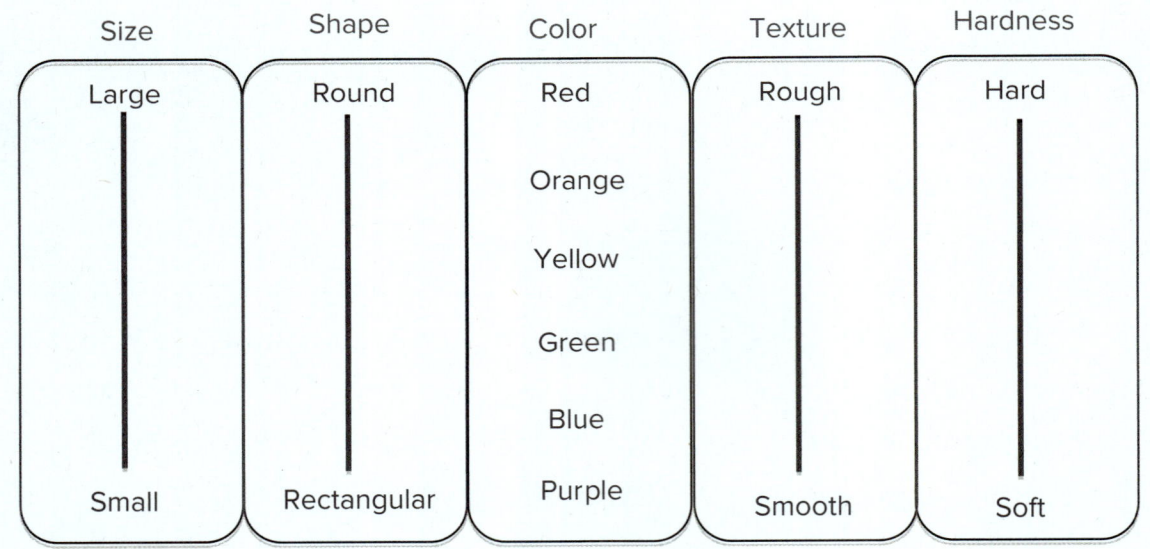

Activity 5
Analyze Like a Scientist

Quick Code: us5014s

Matter

Read the text. As you read, **highlight** evidence that you can use to support your response to the Can You Explain? question.

Matter

Matter is anything that has mass and takes up space. The computer you are using is matter. The juice you drink at breakfast is matter. The air you breathe is matter. Even you are matter! All matter is made up of tiny **particles** that are in constant motion. How much the particles are moving determines the **state of matter**. Light and sound are two examples of things that are not matter. Both of these are considered forms of energy.

This orange juice is an example of matter. Can you think of other examples?

Common states of matter are **solid**, **liquid**, and **gas**. In solid matter, the particles are packed tightly and move only a little bit. In liquid matter, the particles have more space, have more energy, and move more freely. In a gas, the particles have a lot of space and energy and move very freely. Matter can change from one state to another, and these changes, such as ice melting into water or water freezing into ice, happen all the time.

CCC Energy and Matter

Concept 1.1: Describing Matter in Words and Numbers | 17

Matter *cont'd*

All matter can be measured and observed. For example, you can measure how tall you are with a meterstick or a measuring tape. You can measure how much a puppy weighs using a **scale**. You can observe air filling up a balloon, and you can measure how much the balloon expands as it fills. You can observe milk being poured into a glass and measure the amount and temperature of that milk.

The ice cream in this picture is changing state. What is happening to the particles that make up the ice cream?

 **Activity 6**

Observe Like a Scientist

States of Matter

Watch the video. **Look** for ways to describe matter and each of the states of matter.

Quick Code: us5015s

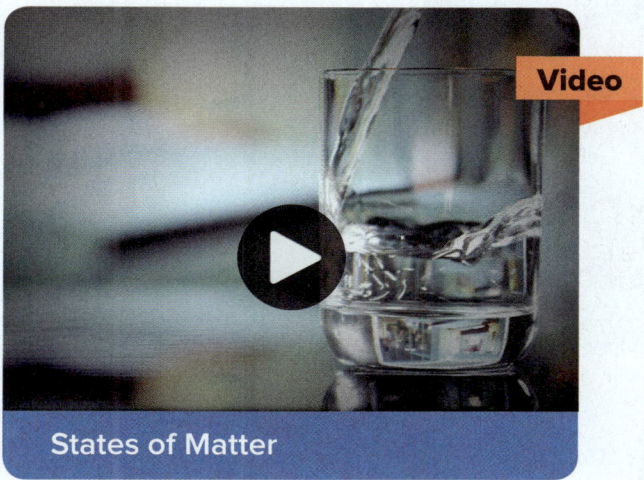

States of Matter

 **Talk Together**

Now, talk together about how you could define *matter*. What is matter? What makes one state of matter different from the others?

CCC Energy and Matter

Concept 1.1: Describing Matter in Words and Numbers | 19

1.1 | Learn How is matter described, measured, and classified?

Activity 7
Evaluate Like a Scientist

What Form Is It?

For each object, **write** its state of matter below the object. Your answers will be *solid*, *liquid*, or *gas*.

Quick Code: us5016s

Ice Cubes

_____

Lava

_____

Smoke

_____

SEP Engaging in Argument from Evidence
CCC Energy and Matter

Wood

Milk

Salt

Mist from a Geyser

_____

_____

_____

_____

Concept 1.1: Describing Matter in Words and Numbers

1.1 | Learn — How is matter described, measured, and classified?

Activity 8

Observe Like a Scientist

Three States of Matter

Quick Code: us5017s

Determine whether each object is a solid, liquid, or gas. **Write** the name of each object in the column where it belongs.

gasoline   wind   apple   orange juice   sand   water   aroma of vinegar

Solid	Liquid	Gas

Use the interactive Three States of Matter to check your thinking.

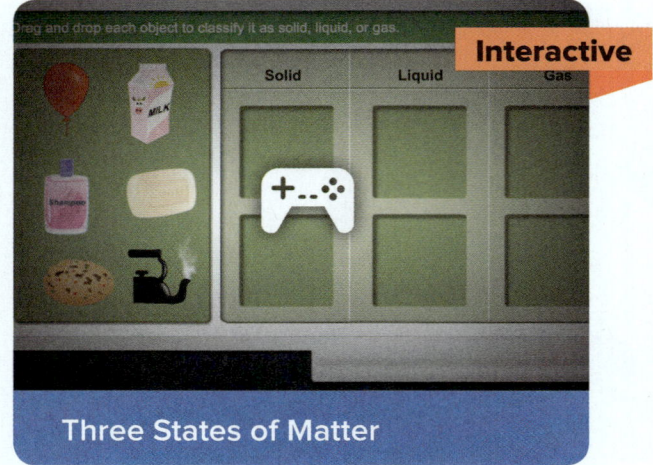

Three States of Matter

SEP Constructing Explanations and Designing Solutions

Look at the rubber band and the eraser. **Identify** the state of matter for each object. **Reflect** on your learning. For the rubber band, **write** an idea that *stretched* your thinking. For the eraser, **write** an old idea that is now *erased* by what you have learned.

Object	State of Matter	Idea That Stretched My Thinking

Object	State of Matter	Idea That Was Erased

Concept 1.1: Describing Matter in Words and Numbers

1.1 | Learn How is matter described, measured, and classified?

How Can the Properties of Matter Be Used to Describe It?

Activity 9

Investigate Like a Scientist

Quick Code: us5018s

Hands-On Investigation: Measuring Properties

In this investigation, you will measure, collect, and analyze properties of various common objects using tools such as rulers, balances, and magnets.

Make a Prediction

If you cut an object in half, how does the mass of one of the pieces compare to the mass of the original object?

_____

_____

_____

What do you think makes an object float?

_____

_____

_____

SEP Planning and Carrying Out Investigations

What materials do you need? (per group)

- Bar magnets
- Balance, triple beam
- Metric ruler
- Beaker, glass, 150 mL
- Paper clips
- Beads
- Aluminum foil
- Wooden blocks
- Water

What Will You Do?

1. Choose properties to investigate that you can observe or measure.
2. Determine the tools needed to investigate each property.
3. Describe the objects using as many properties as possible, including drawing a picture of your object.
4. Make measurements using the tools you chose to use.
5. Test whether the object sinks or floats in a beaker of water.
6. Change the size of one object by breaking, cutting, or tearing the object in half. Observe, measure, and test the object again.
7. Sort your objects into groups.
8. Record what you observed below.

Which properties did you study?

_____

_____

_____

Concept 1.1: Describing Matter in Words and Numbers

1.1 | Learn — How is matter described, measured, and classified?

Write the type of object in the first row. Use the first column to **list** the properties you observed, measured, or tested. **Add** rows to the table as needed.

Type of Object and Its Picture:			
Property 1:			
Property 2:			
Property 3:			
Property 4:			
Property 5:			

Think About the Activity

What tools did you select for this investigation?

_____
_____
_____
_____

How does changing the size of an object change its physical properties?

_____
_____
_____
_____

Describe one of your groups: the objects you included in your group and the reason you grouped those objects.

_____
_____
_____
_____

Activity 10
Analyze Like a Scientist

Properties of Matter

Quick Code: us5019s

Read the text. Then, **circle** any property that you did not measure in Activity 9. **Go back** to the data you collected in Activity 9. **Add** what you have learned from this text to your descriptions.

Properties of Matter

Matter also has many properties that you can describe. Color, shape, odor, and texture are examples of physical properties that you can observe with your five senses. You can use words such as *rough*, *blue*, *smelly*, *round*, and *sweet* to describe these properties.

This picture shows a burning match. What kind of property is the ability to burn?

The ability to burn and the ability to rust are properties that describe how matter interacts with other matter. These are called chemical properties. An important feature of chemical properties is that they are only measurable by changing the **material**.

SEP Obtaining, Evaluating, and Communicating Information

For example, a chemical **property** of paper is that it is flammable. When paper is lit on fire, it becomes ash.

Volume, mass, and temperature are properties of matter that you can measure. Volume is the amount of space that matter takes up. Scientists measure volume in liters (L), milliliters (mL), or cubic centimeters (cm^3). One liter equals 1000 milliliters or 1000 cubic centimeters (1 L = 1000 mL = 1000 cm^3). A big bottle of soda that you might buy for a party typically holds 2 liters.

Mass is a measure of the amount of matter. Scientists often measure mass in grams (g) or kilograms (kg). A dollar bill has a mass of about 1 gram. One kilogram is equal to 1000 grams (1 kg = 1000 g). One liter of water has a mass of 1 kilogram.

Recall that matter is made up of particles in motion. Temperature is a measure of how quickly the particles in a **substance** are moving. Quickly moving particles can give off more heat energy than slower moving particles.

1.1 | Learn — How is matter described, measured, and classified?

Activity 11
Evaluate Like a Scientist

Quick Code: us5020s

Measuring Matter

Shar measured several materials. Her measurements are in the table. Note that temperature is measured in degrees Celsius (°C), mass is measured in grams (g), and volume is measured in milliliters (mL). **Examine** the data in the table.

	Mass (g)	Temperature (°C)	Volume (mL)
Material 1	189	37	100
Material 2	150	55	115
Material 3	99	23	5

Based on the data in the table, **circle** the correct words to make each statement true.

[Material 1/Material 3] contains more matter than Material 2.

[Material 2/Material 3] is cooler than Material 1.

[Material 2/Material 3] takes up more space than Material 1.

SEP Analyzing and Interpreting Data

How Can the Unique Properties of Matter Be Useful?

 **Activity 12**
Analyze Like a Scientist

Quick Code: us5021s

Useful Properties of Matter

Read the text. **Circle** each type of material. **Underline** the properties of that material. Then, **complete** the activity that follows.

Useful Properties of Matter

Helium is a gas that is used to fill balloons. Its properties make it useful for this purpose. For example, a balloon filled with helium gas is lighter than air, so balloons filled with helium float in the air. This is a physical property of helium. Also, helium is not poisonous or flammable, so it is safe to use. (A flammable material is easily set on fire.) Both of these are examples of chemical properties.

These balloons are filled with helium gas. What property of helium makes it useful for filling party balloons?

SEP Obtaining, Evaluating, and Communicating Information
CCC Structure and Function

Concept 1.1: Describing Matter in Words and Numbers | 31

Useful Properties of Matter *cont'd*

Copper is a metal used to make electrical wires. Its physical properties make it useful for this purpose. Copper can be stretched into a thin, flexible wire, which is a physical property. Copper also conducts electricity well, which is another physical property. In contrast, it would not be useful to make wires out of a material like wood. Unlike copper, wood cannot be easily stretched and does not conduct electricity well.

Glass is used to make windows and lightbulbs. What properties of glass make it useful for these purposes?

Diamonds are one of the hardest materials on Earth. They are so hard that they can be used to drill through rock. How do the properties of a diamond determine how they are used?

For each object, **write** an additional application.

Material	Additional Application of the Material
Helium	
Copper	
Glass	
Diamonds	

Activity 13
Evaluate Like a Scientist

Uses of Matter

Circle the properties that make each type of material useful for its purpose.

Type of Matter	Purpose	Properties
Steel	Tools such as screwdrivers and hammers	hard flexible strong flammable liquid waterproof
Oil	Gasoline	hard flexible strong flammable liquid waterproof
Rubber	Tires, wetsuits, gloves	hard flexible strong flammable liquid waterproof

SEP Constructing Explanations and Designing Solutions
CCC Structure and Function

1.1 | Learn — How is matter described, measured, and classified?

Activity 14

Investigate Like a Scientist

Quick Code: us5023s

Hands-On Investigation: Temperature Rising

In this investigation, you will measure the temperature increase on the outside of different containers after warm water is poured into the containers. You will determine which materials would make good insulators and which materials would make good conductors.

Make a Prediction

Which material do you think is a good conductor? Which material do you think is a good insulator?

_____

_____

_____

_____

SEP Planning and Carrying Out Investigations

SEP Analyzing and Interpreting Data

What materials do you need? (per group)

- Safety goggles (per student)
- Thermometer, plastic
- Rubber bands
- Water
- Stopwatch
- Foam cup, 6 oz
- Paper cup, 360 mL
- Plastic cup, 9 oz
- Beaker, glass, 150 mL
- Can, metal

What Will You Do?

1. Place a foam container on a tabletop surface, and then use rubber bands to secure a thermometer to the outside of the container. Make sure that the thermometer is not touching the tabletop surface and is positioned below the water line inside the container.

2. Record the temperature before water is added to the container.

3. Next, add warm water to the container so that it is close to the top but not overflowing.

4. Record the temperature of the container at 5-second intervals until the temperature stops changing or for 2 minutes, whichever comes first.

5. Follow cleanup instructions the teacher provides.

6. Repeat steps 1–5 using a paper container, a plastic container, a glass container, and a metal container.

7. Compare your results to those of other groups.

Concept 1.1: Describing Matter in Words and Numbers

1.1 | Learn How is matter described, measured, and classified?

Time (min:sec)	Foam Container Temperature (°C)	Paper Container Temperature (°C)	Plastic Container Temperature (°C)	Glass Container Temperature (°C)	Metal Container Temperature (°C)
Prior to water addition					
0:00 after water addition					
0:05					
0:10					
0:15					
0:20					
0:25					
0:30					
0:35					
0:40					
0:45					
0:50					
0:55					
1:00					
1:05					
1:10					
1:15					
1:20					
1:25					
1:30					
1:35					
1:40					
1:45					
1:50					
1:55					
2:00					

Think About the Activity

Which material had the fastest increase in temperature?

_____
_____
_____
_____
_____

Which material had the slowest rise in temperature?

_____
_____
_____
_____
_____

How did the size, shape, or thickness of the container affect the results?

_____
_____
_____
_____
_____

1.1 | Learn How is matter described, measured, and classified?

Which material would be best to design a container that keeps hot soup warm for the longest amount of time? Why?

_____
_____
_____
_____

Which material would be best to design a container that keeps a cold drink cool for the longest amount of time? Why?

_____
_____
_____
_____

Which material would be best to design a container that cooks food? Why?

_____
_____
_____
_____

1.1 | Share — How is matter described, measured, and classified?

Activity 15
Record Evidence Like a Scientist

Hands and Hot Chocolate

Quick Code: us5024s

Now that you have learned about properties of matter, you will investigate how a person can hold a cup of hot chocolate without being burned. You first saw this in Wonder.

Let's Investigate Hands and Hot Chocolate

Talk Together

How can you describe Hands and Hot Chocolate now? How is your explanation different from before?

SEP Constructing Explanations and Designing Solutions

Now, **read** the text. **Look for** the claim, evidence, and reasoning in this text.

> The cup of hot chocolate feels good to the person holding the cup. The way the person is holding the cup indicates she feels some warmth, but not so much warmth that she can't safely hold on to the cup. This indicates that the material the cup is made of allows some heat to pass through it, but not so much that the cup is too hot to hold. The ability of the material to pass some heat, but not too much, is considered a property of the material.

A scientific explanation consists of a claim, evidence to support the claim, and reasoning connecting the evidence to the claim. The **claim** is that the cup of hot chocolate feels good to the person holding the cup. The **evidence** is the way in which the person is comfortably holding the cup. The **reasoning** is that the way the person is comfortably holding the cup implies the cup can let some heat pass from the hot chocolate through the cup to the person's hands, but not too much.

Look again at the Can You Explain? question. You first read this question at the beginning of the lesson.

Can You Explain?

How is matter described, measured, and classified?

How can this explanation help you answer the Can You Explain? question or one of your own questions?

My Question

Concept 1.1: Describing Matter in Words and Numbers | 41

1.1 | Share How is matter described, measured, and classified?

Write your scientific explanation.

Claim

Evidence

Reasoning

STEM in Action

Activity 16
Analyze Like a Scientist

Quick Code: us5025s

Careers and Measuring Matter

Read the text. **Complete** the activity that follows.

Careers and Measuring Matter

Understanding and measuring matter are essential to many jobs. Architects and builders use careful measurements of materials when constructing homes and schools. Builders must know correct lengths and widths of boards before putting up walls. Architects need to understand properties of materials, like strength and durability. Knowledge of properties and correct measurements help ensure buildings are safe.

Bakers constantly measure the volume and weight of ingredients. In recipes, the ingredient amount must be precise. For example, too much or too little baking powder can ruin a cake. The correct ratio of dry and wet ingredients gives the right texture to baked goods.

SEP Obtaining, Evaluating, and Communicating Information

Concept 1.1: Describing Matter in Words and Numbers

Careers and Measuring Matter *cont'd*

Measuring Young Trout

Scientists often measure matter during their research. Paleontologists measure the size and shape of fossils. Space scientists measure the mass of planets and stars. Marine biologists measure the speed and volume of sound from animals like whales and dolphins.

Cartographers make maps using technology to measure landforms. They measure the heights of mountains using a telescopic tool called a theodolite. They use an echo sounder that detects timed, sonic pulses to measure the depths of the sea. Satellites help collect data on large areas of Earth. Cartographers can use these measurements to make detailed maps.

Cartographers measure and combine many different properties of landforms to create maps. How are these map measurements used in everyday life? How can they help in emergency situations?

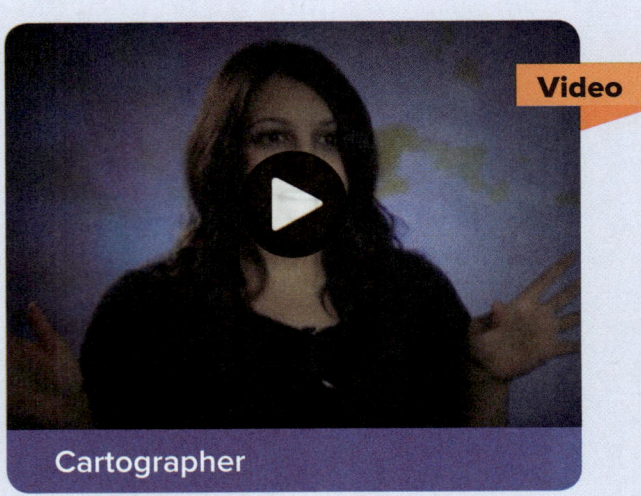
Cartographer

Measuring Steel

This picture shows a bridge made of steel framing.

What are three properties of the steel beams that should be measured either before or during construction of the bridge? Explain why it is important to make each type of measurement.

_____
_____
_____
_____
_____
_____
_____
_____

1.1 | Share — How is matter described, measured, and classified?

Activity 17

Evaluate Like a Scientist

Quick Code: us5026s

Review: Describing Matter in Words and Numbers

Think about what you have read and seen. What did you learn?

Write down some core ideas you have learned. **Review** your notes with a partner. Your teacher may also have you take a practice test.

_____
_____
_____
_____
_____
_____
_____
_____
_____
_____
_____
_____
_____
_____
_____
_____

SEP Obtaining, Evaluating, and Communicating Information

 ## Talk Together

Think about what you saw in Get Started. Use your new ideas about describing matter to discuss water disappearing from a fishbowl.

Concept 1.1: Describing Matter in Words and Numbers | 47

CONCEPT
1.2

Changes to Matter

Student Objectives

By the end of this lesson:

- ☐ I can investigate whether mixing two or more substances makes a new substance.
- ☐ I can identify causes of changes in the physical and chemical properties of matter.
- ☐ I can collect and graph data to provide evidence of what happens when matter changes form.
- ☐ I can plan and conduct an investigation to collect data that describe the effect of heat on the state of matter.

Key Vocabulary

- ☐ atmosphere
- ☐ boil
- ☐ change of state
- ☐ chemical change
- ☐ combine
- ☐ melt
- ☐ mixture
- ☐ physical change
- ☐ plasma
- ☐ sugar
- ☐ water

Quick Code: us5028s

Concept 1.2: Changes to Matter | 49

Activity 1

Can You Explain?

What happens to the mass of a substance when it is heated, cooled, or mixed with other substances?

_____
_____
_____
_____
_____
_____
_____
_____
_____
_____
_____

Quick Code:
us5029s

1.2 | Wonder
What happens to the mass of a substance when it is heated, cooled, or mixed with other substances?

Activity 2
Ask Questions Like a Scientist

Melting Matter

Quick Code: us5030s

Watch an ice cube melting, or **watch** the video. Then, **write** three of your own Wonder questions.

Let's Investigate Melting Matter

SEP Asking Questions and Defining Problems
CCC Cause and Effect

Write three questions about a melting ice cube.

I wonder...

I wonder...

I wonder...

1.2 | Wonder

What happens to the mass of a substance when it is heated, cooled, or mixed with other substances?

Activity 3

Observe Like a Scientist

Quick Code: us5031s

Exploring Solids, Liquids, and Gases

Watch the video. **Look** for characteristics of solids, liquids, and gases.

Exploring Solids, Liquids, and Gases

Talk Together

Now, talk together about the three forms of matter. Can you identify some solids, liquids, and gases in the world around you?

Three Forms of Matter

Write three true statements about the three forms of matter. **Write** one false statement about the three forms of matter, from the video. **Ask** a partner to identify which statement is incorrect.

1.2 | Wonder

What happens to the mass of a substance when it is heated, cooled, or mixed with other substances?

Activity 4
Evaluate Like a Scientist

What Do You Already Know About Changes to Matter?

Quick Code: us5032s

Which States of Matter Do You Recognize?

Look at the three pictures below. **Draw** a line to the correct state of matter.

Air in Balloon

Icicles

Water

Liquid

Solid

Gas

Describing the Three States of Matter

What are some properties of an ice cube that tell you it is a solid? **Write** at least one property, and **explain** how it relates to an ice cube.

_____

_____

_____

Changes in Matter

Does the amount of matter change during a state change? **Circle** the best response to fill in the blank to complete the sentence below.

> stays the same   increases   decreases

When matter changes state, the total number of particles in the matter _____.

1.2 | Learn
What happens to the mass of a substance when it is heated, cooled, or mixed with other substances?

How Does Temperature Affect the State of Matter?

Activity 5
Investigate Like a Scientist

Quick Code: us5033s

Hands-On Investigation: Changing States of Matter

In this investigation, you will investigate the effect of cooling and heating on states of matter.

Make a Prediction

Why does ice melt? What happens to ice when it melts?

_____

_____

_____

Safety Note

- Follow all lab safety guidelines.
- Do not eat or drink anything in the lab.
- Be careful when touching the cooled and heated objects.

SEP Planning and Carrying Out Investigations
CCC Stability and Change

What materials do you need? (per group)

- Freezer
- Hairdryer(s)
- Plates or saucers
- Bouncy balls
- Rubber bands
- Chocolate bars
- Books
- Bar soap
- Water
- Bubbles
- Milk
- Honey
- Balloons
- Plastic zipper bags
- Modeling clay
- Soda
- Dish soap
- Pencils
- Plastic container, 12 oz

What Will You Do?

1. Identify safety concerns.

2. On day 1, your teacher will show you many different objects before placing them in the freezer overnight. Write the name of each object in the data table. Predict what will happen to the object after being frozen. Record your prediction in the data table.

3. On day 2, your teacher will show you the same objects after they have been frozen. Write down your observations in the data table.

4. Next, you will make predictions about what will happen when each object is heated with a hairdryer. Record your predictions in the data table.

5. Follow your teacher's instructions on how to heat the objects. Record your observations in the data table. Clean up your materials as instructed by your teacher.

Concept 1.2: Changes to Matter

1.2 | Learn

What happens to the mass of a substance when it is heated, cooled, or mixed with other substances?

Object	Freezer Prediction	Freezer Observation

Object	Hairdryer Prediction	Hairdryer Observation

1.2 | Learn
What happens to the mass of a substance when it is heated, cooled, or mixed with other substances?

Think About the Activity

Did the cooled items change as you predicted they would? If so, how did you know they would change that way? If not, why was your prediction incorrect?

_____
_____
_____
_____

Did the heated items change as you predicted they would? If so, how did you know they would change that way? If not, why was your prediction incorrect?

_____
_____
_____
_____

What happened to the gases when their containers were cooled?

_____
_____
_____

What happened to the gases when their containers were heated?

_____
_____

Activity 6
Analyze Like a Scientist

Quick Code: us5034s

Temperature and State of Matter

Read the text and **watch** the video. **Highlight** evidence you can use to answer the Can You Explain? question or one of the questions you wrote in the Wonder section.

Temperature and State of Matter

A substance's state depends partly on its temperature. A substance's temperature is a measure of how much energy the particles in that substance have. It is the energy of the particles that determine how much they move and, therefore, the state of the matter.

Changes of State

SEP Constructing Explanations and Designing Solutions
CCC Cause and Effect
CCC Energy and Matter

Concept 1.2: Changes to Matter | 63

Temperature and State of Matter cont'd

For example, **water** is a liquid between 0°C and 100°C. Water becomes a solid when it is cooled below 0°C, its freezing point; its state changes from liquid to solid. As the particles of liquid water lose energy, they slow down until the liquid water becomes solid ice.

Melting is the opposite process. Melting is the **change of state** from solid to liquid. It happens when energy is added to a solid. For example, as particles of solid ice gain energy, they move around more. Eventually, they move around enough that the water begins to flow, or **melt**. This also happens at 0°C.

These frozen treats are changing state. Why is this happening?

Changes of state are physical changes often caused by changes in temperature. Physical changes do not change the makeup of a substance, and they are usually reversible. For example, melting is a **physical change** that can be reversed by cooling liquid water until it freezes again. The water is still water—it is the same substance—whether it is liquid or solid, even though it looks different. Increasing or decreasing temperature can also cause chemical changes.

Choose one of the objects you used in the Changing States of Matter activity. Fill in the Change over Time graphic organizer below. On the Topic line, **write** the name of your object and the treatment the object received (heating or cooling). **Draw a model** of the object before you applied the treatment. **Draw a model** of the object after you applied the treatment. In the box at the bottom, **write** an explanation for the changes you observed.

Topic: _____

Before:	After:

Changes:

Concept 1.2: Changes to Matter

1.2 | Learn
What happens to the mass of a substance when it is heated, cooled, or mixed with other substances?

How Are Different Types of Mixtures Formed?

Activity 7
Observe Like a Scientist

Real-World Mixtures

Quick Code: us5035s

Look at the three images. Then, **answer** the question.

1: Pink Granite

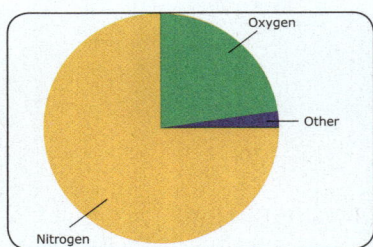

2: Atmosphere

3: Ocean Water

Which matches your definition of a mixture? **Describe** the parts of the mixture.

SEP Constructing Explanations and Designing Solutions
CCC Structure and Function

Activity 8
Analyze Like a Scientist

Mixtures

Read the text. **Highlight** characteristics of mixtures. **Think** about the particle size of the mixtures mentioned in the text.

Quick Code: us5036s

Mixtures

A **mixture** is a form of matter made of two or more parts. Each part in a mixture keeps its own identity—in other words, mixing the parts does not change them into new substances. A mixture can be made of solids, such as a mixture of nuts and bolts. Or it can include a combination of a solid and a liquid, such as a mixture of salt and water. The **atmosphere** is a mixture of many gases.

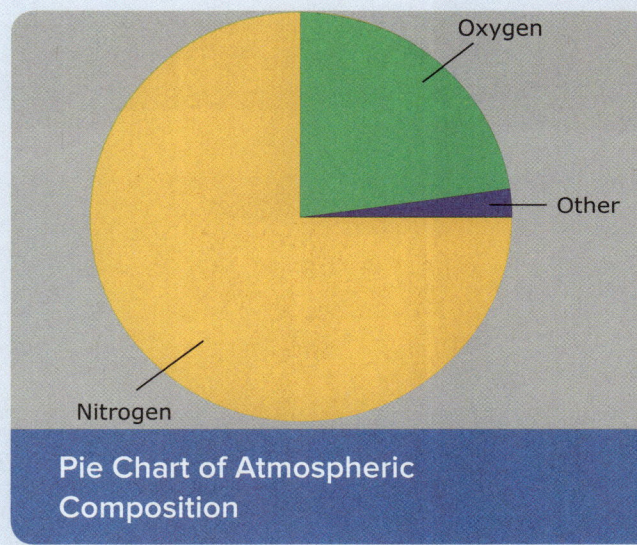

Pie Chart of Atmospheric Composition

Compare a mixture of nuts and bolts with a mixture of gases. Both are types of mixtures, and both have different parts. But you can easily see the different parts in the mixture of nuts and bolts. You would need special equipment to see the parts in a mixture of gases.

SEP Obtaining, Evaluating, and Communicating Information
CCC Scale, Proportion, and Quantity

Concept 1.2: Changes to Matter | 67

Mixtures cont'd

This pile of nuts is a mixture. Can you think of some other common mixtures that you encounter in your daily life?

Water Filter

When materials are mixed and form a mixture, they do not **combine** chemically. Each material keeps the properties that you can use to identify it. For example, **sugar** does not lose its sweetness when it is mixed with water. This also means that you can separate the parts of a mixture. There are different methods to separate mixtures. A filter can separate mixtures if one material has smaller particles than the other. Evaporation can separate some mixtures because the materials will evaporate at different temperatures.

Draw a diagram using circles of different sizes to represent the different particle sizes in the three different mixtures in this text.

Nuts and Bolts	Atmosphere	Sugar and Water

Concept 1.2: Changes to Matter | 69

1.2 | Learn
What happens to the mass of a substance when it is heated, cooled, or mixed with other substances?

Activity 9

Investigate Like a Scientist

Quick Code: us5037s

Hands-On Investigation: Mixing It Up

In this investigation, you will explore what happens when you mix substances together.

Make a Prediction

What is the question you want to answer? Write the question for the investigation. The question should be specific and investigable.

_____
_____

What do you predict will be the result of the investigation? Develop a claim about what you think is going to happen.

_____
_____

How will you investigate the question? Describe the plan that you will use to study your question and analyze your hypothesis.

_____
_____
_____
_____
_____

SEP Constructing Explanations and Designing Solutions
CCC Patterns

What materials do you need? (per group)

- Plastic zipper bags
- Pocket scale
- Spoon, plastic
- Weighing dishes
- Safety goggles (per student)
- Disposable gloves (per student)
- Baking soda (sodium bicarbonate)
- Washing soda (sodium carbonate)
- Flour
- Calcium chloride solid, $CaCl_2$
- Cornstarch
- Epsom salts
- Water
- Vinegar
- Lemon juice
- Tincture of iodine
- Juice from purple cabbage
- Powdered lemonade

What Will You Do?

Part 1: Mixing Solids:

1. Choose two solids. Ask your teacher to confirm your choices.

2. Place the weighing dish on the scale, and set the scale to read 0.0 grams with the empty weighing dish on the pan. Add 1 gram of Solid 1 to the weighing dish. Record the mass, and set aside.

3. Place a new weighing dish on the scale, and set the scale to read 0.0 grams with the empty weighing dish on the pan. Add 1 gram of Solid 2 to the weighing dish. Record the mass, and set aside.

4. Find the mass of a plastic zipper bag, and record it.

5. Add Solid 1 and Solid 2 to the zipper bag, and close the bag.

Concept 1.2: Changes to Matter | 71

1.2 | Learn — What happens to the mass of a substance when it is heated, cooled, or mixed with other substances?

6. Mix the two solids with your hands by massaging the zipper bag from the outside, and record your observations.

7. Find the mass of the zipper bag that contains the two solids, and record it.

8. Repeat steps 1–7 for two different solids for as many trials as your teacher requests.

Part 2: Mixing Liquids

1. Choose two liquids. Ask your teacher to confirm your choices.

2. Place the weighing dish on the scale, and set the scale to read 0.0 grams with the empty weighing dish on the pan. Add 1 gram of Liquid 1 to the weighing dish. Record the mass, and set aside.

3. Place a new weighing dish on the scale, and set the scale to read 0.0 grams with the empty weighing dish on the pan. Add 1 gram of Liquid 1 to the weighing dish. Record the mass, and set aside.

4. Find the mass of a plastic zipper bag, and record it.

5. Add Liquid 1 and Liquid 2 to the zipper bag, and close the bag.

6. Mix the two liquids with your hands by massaging the zipper bag from the outside, and record your observations.

7. Find the mass of the zipper bag that contains the two liquids, and record it.

8. Repeat steps 1–7 for two different liquids for as many trials as your teacher requests.

Part 3: Mixing Solids and Liquids

1. Choose a solid and a liquid. Ask your teacher to confirm your choices.

2. Place the weighing dish on the scale, and set the scale to read 0.0 grams with the empty weighing dish on the pan. Add 1 gram of the solid to the weighing dish. Record the mass, and set aside.

3. Place a new weighing dish on the scale, and set the scale to read 0.0 grams with the empty weighing dish on the pan. Add 1 gram of the liquid to the weighing dish. Record the mass, and set aside.

4. Find the mass of a plastic zipper bag, and record it.

5. Add the solid and the liquid to the zipper bag, and close the bag.

6. Mix the solid and the liquid with your hands by massaging the zipper bag from the outside, and record your observations.

7. Find the mass of the zipper bag that contains the two liquids, and record it.

8. Repeat steps 1–7 for two different liquids for as many trials as your teacher requests.

Record your data from your investigation.

Trial 1

Mixture	Substances	Mass before Mixed	Mass after Mixed
Solids	1		
	2		
Liquids	1		
	2		
Solids and Liquids	1		
	2		

Concept 1.2: Changes to Matter | 73

1.2 | Learn
What happens to the mass of a substance when it is heated, cooled, or mixed with other substances?

Trial 2

Mixture	Substances	Mass before Mixed	Mass after Mixed
Solids	1		
	2		
Liquids	1		
	2		
Solids and Liquids	1		
	2		

Trial 3

Mixture	Substances	Mass before Mixed	Mass after Mixed
Solids	1		
	2		
Liquids	1		
	2		
Solids and Liquids	1		
	2		

Trial 4

Mixture	Substances	Mass before Mixed	Mass after Mixed
Solids	1		
	2		
Liquids	1		
	2		
Solids and Liquids	1		
	2		

Think About the Activity

What did you learn from this investigation? Develop a conclusion for your investigation.

_____
_____
_____
_____
_____

Concept 1.2: Changes to Matter

1.2 | Learn

What happens to the mass of a substance when it is heated, cooled, or mixed with other substances?

What happened to the properties of the substances when they were mixed?

_____
_____
_____
_____
_____
_____

What did you observe regarding the mass before and after mixing?

_____
_____
_____
_____
_____
_____

Did you observe any color changes? What does this tell you about the mixture?

_____
_____
_____
_____
_____
_____
_____

Did you observe any gases being formed? What does this tell you about the mixture?

_____
_____
_____
_____

What would have happened to the masses if you had mixed the substances in plastic cups?

_____
_____
_____
_____
_____

What patterns do you observe in the data collected in this activity?

_____
_____
_____
_____
_____

1.2 | Learn
What happens to the mass of a substance when it is heated, cooled, or mixed with other substances?

Activity 10
Evaluate Like a Scientist

Properties of Mixtures

Quick Code: us5038s

Think about the definition of *mixture*. Which of the following properties do all mixtures have? **Circle** the letters for all the choices that apply.

A. made of parts that can be separated

B. are considered pure substances

C. made of parts that react chemically with each other

D. are formed by physically combining two or more substances

E. made of parts that cannot be separated

F. can be liquids, gases, or solids

SEP Constructing Explanations and Designing Solutions

Provide examples that support your answer to the previous question.

_____
_____
_____
_____
_____
_____
_____

How Is a Chemical Change Different from a Physical Change?

Activity 11
Analyze Like a Scientist

Physical and Chemical Changes

Quick Code: us5039s

Read the text. **Look** for the types of changes. Then, **answer** the questions that follow.

Physical and Chemical Changes

Matter can be changed physically or chemically. A physical change affects only the physical properties of a substance. It does not change the substance's chemical identity or produce new kinds of substances. Shaping a piece of clay into a ball is a physical change. The clay is still clay after you change its shape. Freezing and boiling water are both physical changes. Ice is water in its frozen, or solid, state. Water vapor is water in its gas state. Sugar dissolving into water is also a physical change. The sugar molecules spread out through the water molecules, but the sugar is still sugar, and the water is still water. Most physical changes can be reversed easily. For example, if you heat sugar water, the water will evaporate into a gas, while the sugar will remain a solid.

SEP Constructing Explanations and Designing Solutions

A **chemical change** produces a new kind of substance. The new substance is different physically from the original substance. However, it also has different chemical properties. For example, the elements iron and oxygen combine to form rust. Rust is a flaky, reddish chemical called iron oxide. When oxygen combines with carbon and hydrogen, however, they release heat that can start a fire. The fire can change a substance such as wood into ash. Unlike physical changes, chemical changes are not reversed easily.

If it's cold enough for a long enough period of time, even a large river can freeze over. How deep the ice gets depends on the river's current and the length of the cold snap.

What physical changes were described in the text?
_____
_____

What chemical changes were described in the text?
_____
_____
_____

Concept 1.2: Changes to Matter

1.2 | Learn
What happens to the mass of a substance when it is heated, cooled, or mixed with other substances?

Activity 12
Observe Like a Scientist

Chemical Changes in Matter

Watch the video. **Look** for clues that a change in matter is chemical.

Quick Code: us5040s

Chemical Changes in Matter

Now, **complete** the Chemical Changes portion of the interactive. Then, **answer** the questions.

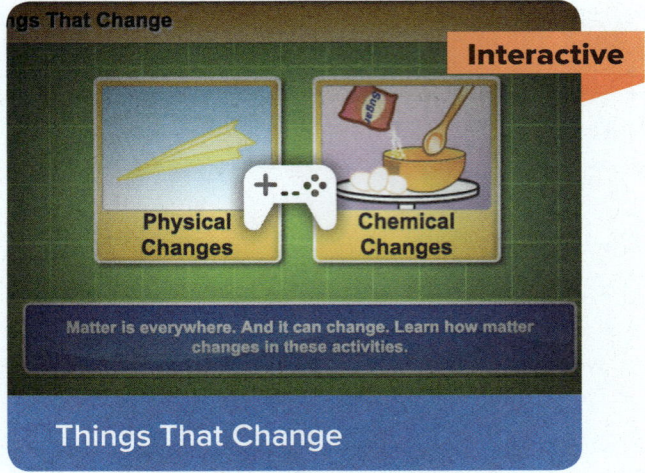

Things That Change

SEP Obtaining, Evaluating, and Communicating Information

82

Which of the activities in the table is a chemical change? **Write** yes or no in the table. For each row, **write** an answer to the question *How do you know?*

Activity	Chemical Change? (yes or no)	How Do You Know?
Make ice cubes		
Bake bread		
Comb your hair		
Burn a candle		
Dye your hair		
Spill bleach on jeans		
Digest your food		
Chew your food		

Concept 1.2: Changes to Matter | 83

1.2 | Learn
What happens to the mass of a substance when it is heated, cooled, or mixed with other substances?

Activity 13

Investigate Like a Scientist

Quick Code: us5041s

Hands-On Investigation: How Has It Changed?

In this investigation, you will explore the difference between physical and chemical changes.

Make a Prediction

How can you tell whether a change is chemical or physical?

_____
_____
_____
_____
_____
_____
_____
_____
_____
_____

SEP Analyzing and Interpreting Data

What materials do you need? (per group)

- Beaker, glass, 150 mL
- Water
- Salt
- Balance, triple beam
- Spoon, plastic
- Hot plate
- Assortment of objects that have experienced various physical changes compared to a standard object (for example, several pieces of ice compared to liquid water, several pretzel sticks broken into different-sized pieces compared to a whole pretzel stick)
- One object that has gone through a chemical change (for example, a rusted nail, an overripe banana)
- Safety goggles (per student)

What Will You Do?

1. Add 100 milliliters of water to a beaker.
2. Measure 20 grams of salt using a triple beam balance. Be sure the 100-gram and 1-gram sliders are all the way to the left.
3. Add the measured salt to the water in the beaker. Then, stir gently with a spoon or a stirring rod.
4. Observe what happens, and record your observations in the Data section.
5. Set aside the beaker for later use.

1.2 | Learn
What happens to the mass of a substance when it is heated, cooled, or mixed with other substances?

6. Get an assortment of objects from your teacher that are made of the same material but have undergone physical changes. Be sure you have one object to use as a standard reference (no changes) and one that has gone through a chemical change.

7. Identify whether each object went through a physical change compared to the standard object. Record the properties that changed, including units where appropriate.

8. Share your results with other groups.

9. Place your beaker with salt water on a hot plate, and heat the mixture to boiling until all the water evaporates. Turn off the hot plate as soon as the water is gone.

10. Record the appearance of the beaker after all the water boiled away.

Record your observations and data.

Think About the Activity

List at least four examples of physical changes that you observed in this activity.
_____
_____

How is a chemical change different from a physical change?
_____
_____
_____

When water heats up and evaporates, has a physical or chemical change occurred? How do you know?
_____
_____
_____

When salt dissolves in water, has a physical or chemical change occurred? How do you know?
_____
_____
_____

Was your prediction accurate?
_____
_____
_____
_____
_____

Concept 1.2: Changes to Matter

1.2 | Learn — What happens to the mass of a substance when it is heated, cooled, or mixed with other substances?

Activity 14
Evaluate Like a Scientist

Physical Changes in Our Lives

Quick Code: us5042s

Read the text. Then, **highlight** the sentences that describe a physical change to a substance but not a chemical change.

Last weekend, we went camping beside a river. As soon as we arrived, we jumped into the water to swim. When we were finished, we sat in the sun. After a while, our wet clothes became dry again. We were hungry, so we decided it was time to eat. We started a fire and cooked some eggs. We also prepared a salad by chopping lettuce, tomatoes, carrots, and apples. Everything tasted great. Then we took a nap so we could digest the food. As I was sleeping, an accident happened: my cousin stepped on my glasses, causing them to break! Fortunately, we found some tape and taped my glasses back together.

SEP Constructing Explanations and Designing Solutions

For each change, **explain** why it is a physical change and not a chemical change.

How Does Changing the State of a Substance Affect Its Mass?

Activity 15
Analyze Like a Scientist

Quick Code: us5043s

No Change in Mass

Often, it is necessary to break larger systems into components so you can measure parts within the system that contribute to the overall function of the system. As you read the text, **look** for the input, processes, and outputs of the system of a pot of boiling water. Then, **answer** the questions.

No Change in Mass

Imagine that you **boil** a pot of water. Imagine that you weigh the mass of the water before you boil it and then again after you boil it. You discover that the mass of the water is slightly less after it has boiled for a few minutes.

Where did the water go? Did the water suddenly lose mass? No, the total mass of an object or substance does not change, even when it turns from a liquid to a gas. In this example, some of the water changed into a gas and left the pot, so the water in the pot weighs less.

SEP Constructing Explanations and Designing Solutions
CCC Systems and System Models

However, the total mass of both forms of the water is still the same. If the masses of the liquid water and water vapor now in the air were added together, it would be the same as the original mass of the water. During a change of state like this, weight is not gained or lost, but it is conserved. Mass does not change in physical changes or chemical reactions. This is called the conservation of mass. Because it applies everywhere, it is therefore called the law of conservation of mass.

If you boil water, how could you measure the mass of the water vapor?

Watch your teacher light a match. After it is blown out, **watch** the smoke rise from the match. Would you get a different mass if you weighed the match on a balance before and after it burned? Why or why not?

_____
_____
_____
_____

Would you get a difference in mass if you burned the match in a closed jar? Why or why not?

_____
_____
_____
_____
_____

Activity 16
Evaluate Like a Scientist

Plan an Investigation

Jeff is doing an investigation to prove that a sample of matter has the same mass before and after a change in state from solid to liquid. The steps are shown in the bottom row but are out of order. **Number** the steps to show the order they should occur. **Share** your ideas with other students to compare your answers.

Quick Code: us5044s

Number of Step					
Steps	Weigh the beaker with the ice cubes, and note the weight.	Weigh the beaker with the water, and note the weight.	Place three ice cubes in a beaker.	Place the ice cubes in a microwave.	Melt the ice cubes using the microwave.

SEP Planning and Carrying Out Investigations

Concept 1.2: Changes to Matter

1.2 | Share
What happens to the mass of a substance when it is heated, cooled, or mixed with other substances?

Activity 17
Record Evidence Like a Scientist

Melting Matter

Now that you have learned about changes to matter, look again at the video Let's Investigate Melting Matter. You first saw this in Wonder.

Quick Code: us5045s

Let's Investigate Melting Matter

Talk Together

How can you describe melting ice now? How is your explanation different from before?

SEP Constructing Explanations and Designing Solutions

A scientific explanation consists of a claim, evidence to support the claim, and reasoning connecting the evidence to the claim. A scientific explanation of the phenomenon of melting without a change in mass would have the following structure:

The **claim** is that there is no change in mass after melting a sample of ice. The **evidence** is that the mass of the sample of ice is identical to the mass of the water that results from the melting of the ice. The **reasoning** is that since the mass of the liquid water after melting was the same as the mass of the solid ice, we can conclude that no mass was gained or lost during this change of state.

Look again at the Can You Explain? question. You first read this question at the beginning of the lesson.

Can You Explain?

What happens to the mass of a substance when it is heated, cooled, or mixed with other substances?

How can this explanation help you answer the Can You Explain? question or one of your own questions?

My Question

1.2 | Share
What happens to the mass of a substance when it is heated, cooled, or mixed with other substances?

Write your scientific explanation.

Claim

Evidence

Reasoning

STEM in Action

Quick Code: us5046s

Activity 18
Analyze Like a Scientist

Careers and States of Matter

Read the text, and **watch** the videos. **Think** about ways you can use different states of matter on the job.

Careers and States of Matter

You may think of states of matter as something that you only learn about in the classroom. But there is a career you are probably familiar with that relies on the three common states of water every day—a chef! Chefs may not seem like scientists, but they mix things together and create new foods every day.

Shannon Shaffer is a chef, and he knows the importance of science in cooking. By using foods in unique ways, Shannon has developed his own style of cooking. As you watch the video, notice how Shannon uses ice water to make "pearls" of vegetables. Think about times you've seen food cooked. Perhaps you've seen an adult boil some water to make spaghetti? Did you see the steam? That's water as a gas. Maybe there were some frozen vegetables that were added to the meal. Freezing vegetables keeps them fresh and ready to use for longer periods of time. The ice on the outside of the package is water in a solid state. Just by preparing a meal, you might have seen water in all three states: solid, liquid, and gas.

SEP Obtaining, Evaluating, and Communicating Information

Careers and States of Matter *cont'd*

Cool Jobs in Science: Shannon Shaffer

Like Shannon Shaffer, you can experiment with the different states of matter in your kitchen. Think about what happens when you add liquid water to an ice cube tray. If you place the filled tray into the freezer for several hours, what comes out? What if you placed those ice cubes in a hot pan? What kind of changes would you see? (Don't try this without an adult's supervision.)

Can you think of other careers that need to have knowledge of the three common states of matter?

Taste the States of Matter

You have seen that scientists, including chefs, have to be careful about how they measure, classify, and describe things. Imagine that you are a chef, and you want to impress your guests with a special themed dinner called "Taste the States of Matter." You need to plan a creative meal that includes various flavors and illustrates the three main states of matter. What would you prepare for your guests? How would you plan the meal? What safety considerations should you make?

_____
_____
_____
_____
_____
_____
_____
_____
_____
_____
_____
_____
_____
_____
_____

Another Chef's Meal

Suppose another chef created a "Taste the States of Matter" dinner, as shown in the image below.

Pasta with Vegetables

Did this chef create a meal that illustrates the three main states of matter? **Explain** your reasoning.

_____
_____
_____
_____
_____

Activity 19

Evaluate Like a Scientist

Quick Code: us5047s

Review: Changes to Matter

Think about what you have read and seen. What did you learn?

Write down some core ideas you have learned. **Review** your notes with a partner. Your teacher may also have you take a practice test.

_____
_____
_____
_____
_____
_____
_____

Talk Together

Think about what you saw in Get Started. Use your new ideas about changes in matter to discuss water disappearing from a fishbowl.

SEP Obtaining, Evaluating, and Communicating Information

Concept 1.2: Changes to Matter | 101

CONCEPT 1.3

A Model of Matter

Student Objectives

By the end of this lesson:

- ☐ I can develop a model of matter representing the large quantity of particles too small to be seen.
- ☐ I can explain the cause-and-effect relationships among temperature, behavior of particles in matter, and state of matter.

Key Vocabulary

- ☐ atom
- ☐ electron
- ☐ energy
- ☐ heat
- ☐ light
- ☐ model
- ☐ thermal energy
- ☐ warm

Quick Code: us5049s

Concept 1.3: A Model of Matter

Activity 1

Can You Explain?

How can a model show us the behavior of the particles that make up matter?

Quick Code:
us5050s

Concept 1.3: A Model of Matter

1.3 | Wonder

How can a model show us the behavior of the particles that make up matter?

Activity 2
Ask Questions Like a Scientist

Quick Code: us5051s

Models

Watch the video. Then, **complete** the activity that follows.

Let's Investigate Models

SEP Developing and Using Models

Complete the sentences with what you wonder about models.

I wonder _____

I wonder _____

I wonder _____

I wonder _____

Decide which of your statements could be converted to testable questions. For those questions, **write** the variables that could be investigated.

_____

_____

_____

_____

Concept 1.3: A Model of Matter

1.3 | Wonder

How can a model show us the behavior of the particles that make up matter?

Activity 3
Observe Like a Scientist

What Is a Model?

Watch the video. As you watch, **observe** the various models presented.

Quick Code: us5052s

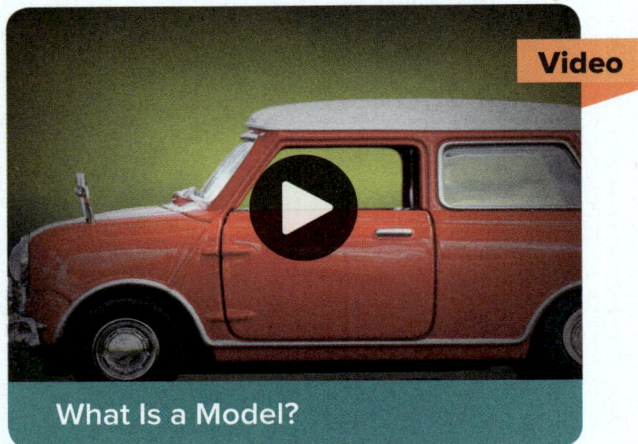

What Is a Model?

Talk Together

Now, talk together about how models are used in real-world applications.

SEP Developing and Using Models

Activity 4
Evaluate Like a Scientist

Quick Code: us5053s

What Do You Already Know About Models of Matter?

How are particles arranged in the different states of matter? **Match** the particle diagrams with the pictures showing different states of matter.

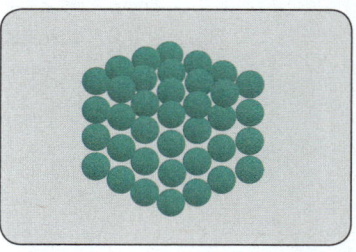

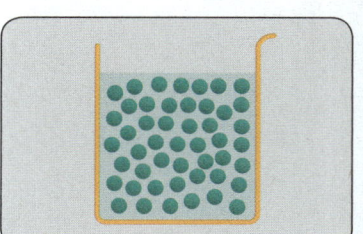

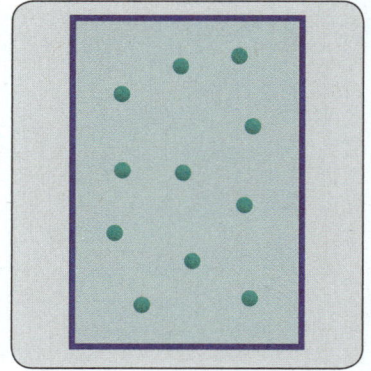

SEP Developing and Using Models

Concept 1.3: A Model of Matter

1.3 | Learn
How can a model show us the behavior of the particles that make up matter?

How Can We Model the Particles That Make Up Matter?

Activity 5

Observe Like a Scientist

What Is Matter?

Watch the video. As you watch, **look for** ways we can collect evidence about the existence of matter.

Quick Code: us5054s

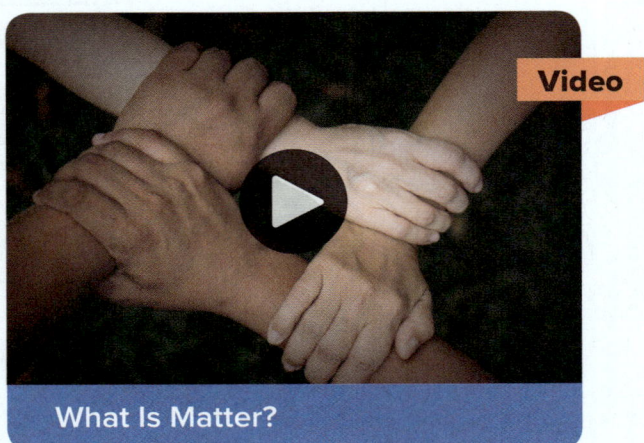

What Is Matter?

Talk Together

Now, talk together about why we cannot see, with our eyes, individual particles that make up matter.

SEP Constructing Explanations and Designing Solutions

Activity 6
Analyze Like a Scientist

Particles of Matter

Quick Code: us5055s

Read the text. As you read, **highlight** evidence to support this claim: Particles are often called "the building blocks of matter."

Particles of Matter

Everything around you, even your body, is made up of matter. We define *matter* as anything that has mass and takes up space. Solids, liquids, and gases are all states of matter. What is matter made of? Imagine what would happen if you could break down a chunk of matter, like a piece of gold, into smaller and smaller pieces. Eventually, the pieces would get so small you could no longer see them, even with a microscope.

Gold Particles

You would end up with extremely small pieces of matter that we'll call particles. There are many different kinds of particles. Different kinds of matter are made of different kinds of these particles.

SEP Developing and Using Models

Concept 1.3: A Model of Matter | 111

1.3 | Learn
How can a model show us the behavior of the particles that make up matter?

Activity 7
Evaluate Like a Scientist

Modeling the Particles of Matter

Quick Code: us5056s

Read the scenario. **Write** or **draw** a note to your friend describing what happened. In your note, use one or more of the following terms: *matter*, *particle*, *solid*, *liquid*, and *gas*.

You and a friend were playing with ice cubes outside on a hot summer day. You both were called away to do a chore and forgot to clean up. Several ice cubes were left on a picnic table outside in the sun. When you returned several hours later, there were no ice cubes or water left on the table, and your friend was puzzled and worried. What happened to the ice cubes?

Your student group is developing a model to show how atoms make up matter. Your job is to choose an object to represent atoms in the model. **Circle** the objects you will choose.

syrup   ping pong balls   tiny pieces of paper   a spectrum (rainbow)

Now, **explain** why you chose the objects you did.

_____
_____
_____

SEP Engaging in Argument from Evidence

How Can We Describe the Particle Size in Matter?

Activity 8
Analyze Like a Scientist

Quick Code: us5057s

Tiny Particle Size

Read the text. As you read, **record** evidence in support of the Can You Explain? question.

Tiny Particle Size

The exact size of a particle depends on the type of particle and how it connects with neighboring particles. The average size of a particle is so tiny that one of your hairs is about 150,000 to 300,000 particles thick!

Scientists can use special microscopes called **electron** microscopes to see individual particles; the microscopes you have in your classroom are not powerful enough to see them. If the tiny size of particles makes them too small to see, even with microscopes, how can we tell they are actually there?

A tiny blood cell can be seen under the high power of a microscope. Each of these blood cells is made up of about 100 trillion particles.

CCC Scale, Proportion, and Quantity

Concept 1.3: A Model of Matter

Tiny Particle Size *cont'd*

Examining gases can help demonstrate that these invisible particles really do exist. Think about what happens when you blow up a balloon. Even though the gas in the balloon is invisible, it still is made up of particles of air. The particles in a gas move very quickly. They bounce against the inside of the balloon. This exerts a force that inflates the balloon and creates its round shape. If you squeeze the balloon, you can make it smaller by pushing the particles closer together. But if you squeeze it too hard, the balloon pops, and the particles that were inside escape into the air.

My Evidence:

___
___
___
___
___

How Does Temperature Affect the Particles in Matter?

 **Activity 9**
Analyze Like a Scientist

Quick Code: us5058s

Particles in Motion

Read the text about Particles in Motion. Then, **discuss** with your class the question that follows.

Imagine you could shrink to the scale of the tiny particles that make up matter and move around in a cup of hot chocolate. **Write** about or **draw** what you would experience.

SEP Developing and Using Models
CCC Energy and Matter
CCC Scale, Proportion, and Quantity

Concept 1.3: A Model of Matter | 115

Particles in Motion

Heat is a form of **energy** you use every day. You heat your hands in front of a fireplace and cook s'mores over a campfire. You use heat to **warm** your home. Heat from the sun keeps living things on Earth alive. Heat is not a physical thing or material, like a cup of hot chocolate. It is simply a form of energy that can make a chocolate drink hot. Heat is also known as **thermal energy**.

Hot Chocolate

Matter is anything that takes up space and has mass. A chocolate drink, like all matter, is made of extremely small particles. These particles have energy. This energy makes the particles move, vibrate, and spin around. When **light** energy or thermal energy is absorbed by matter, the particles in the matter move, vibrate, and spin faster. The faster all this movement is, the more thermal energy the object has. The more thermal energy the object has, the warmer it is to the touch. It's important to remember that the particles that make up matter are always moving in some way.

How can marbles, or other visible particles, act as a model to describe and explain some of the properties and behavior of matter? **Write** or **draw** your ideas.

_____
_____
_____
_____
_____

Activity 10
Evaluate Like a Scientist

Model Particle Motion

Circle the word or phrase from each word bank that best completes each sentence.

Robin flipped a box top over so it was like a tray. She poured marbles into the box top. She used the marbles in the box top to model the motion of particles in an ice pop. First, she poured juice into ice pop molds and placed them in the freezer. After two hours, the ice pops were completely solid. She modeled the state change by moving the box top quickly, then _____.

| stopping | more slowly | even faster |

Next, she took the ice pops out of the freezer. After an hour, they were _____.

| solid | liquid | gas |

She modeled the state change by _____

| decreasing | increasing | stopping |

the motion of the marbles in the box top.

Throughout all parts of this investigation, Robin observed that the _____

| mass | volume | density |

of the ice pops increased when they were frozen.

SEP Developing and Using Models

1.3 | Learn

How can a model show us the behavior of the particles that make up matter?

Now, **draw** the motion of particles in the three different states.

What Is the Relationship between Changing States and How a Particle Moves?

Activity 11
Think Like a Scientist

Modeling States of Matter

Quick Code: us5060s

In this activity, you will develop a model to represent the different states of matter: solid, liquid, and gas.

What materials do you need? (per group)

- Small pompoms, about 40
- Glue
- Index cards, 3, 4 × 6 or larger
- Markers

What Will You Do?

1. Use a marker to label each index card: *solid*, *liquid*, and *gas*.
2. Glue pompoms to index cards to represent models of the states of matter: solid, liquid, and gas.

SEP Developing and Using Models

Concept 1.3: A Model of Matter | 119

1.3 | Learn
How can a model show us the behavior of the particles that make up matter?

Think About the Activity

Describe the arrangement of particles in the different states of matter you modeled in this investigation.

_____
_____
_____
_____
_____

What is matter composed of?

_____
_____
_____

Give examples of solids, liquids, and gases that you use every day.

_____
_____
_____
_____
_____

What does the arrangement of particles in solids, liquids, and gases tell us about the objects we are able to see and not see?

_____

_____

_____

_____

_____

_____

How Can the Particle Model of Matter Help to Explain Everyday Events?

Activity 12
Analyze Like a Scientist

Particles Always in Motion

Quick Code: us5061s

Read the text, and **watch** the videos. Then, **discuss** the questions that follow with your class.

Particles Always in Motion

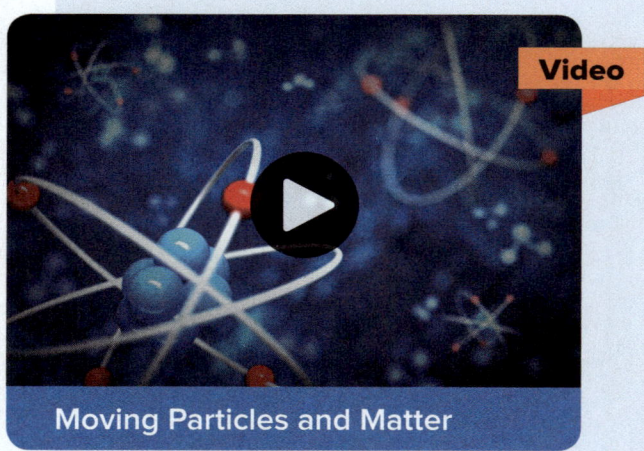
Moving Particles and Matter

We now have a model of matter that says that all matter is made up of extremely tiny particles that are always moving. This will help us begin to understand some of the properties of matter and why it behaves the way it does.

Take a look at the phenomenon of conduction. In the process of conduction, the faster vibrating particles in a warmer object make the slower particles of a cooler

SEP Developing and Using Models
CCC Scale, Proportion, and Quantity
CCC Energy and Matter

object vibrate faster. How do they do this? By bumping into them! In this way, thermal energy is transferred from warm objects to cooler objects by direct contact or from the warmer end of a metal rod to the cooler end.

Conductors and Insulators

The particle model of matter can also explain the water cycle. As water on the ground or in the sea is warmed by the sun, its particles gain energy. The water vaporizes (or evaporates) in the sun and turns into water vapor. This rises into the sky, where it cools and condenses to form the water droplets in clouds. These eventually grow bigger and fall as rain. If it is very cold, they lose energy, freeze, and fall as snow or hail.

 ## Talk Together

How is the description of the bumping of particles into each other another detail of the particle model of matter that describes and explains how conduction in metals works? Does this particle model of matter help you think about another real-world phenomenon?

1.3 | Learn
How can a model show us the behavior of the particles that make up matter?

Activity 13
Evaluate Like a Scientist

Quick Code: us5062s

Modeling Gas

While atoms and molecules are too small to see without powerful microscopes, models can be used to help us understand their structure. A student inflates a balloon and uses it to observe the behavior of a gas. The student is developing a model for the particles of gas in the balloon.

Circle the word that best completes each sentence to describe the model.

The model should show how gas is made up of _____

| smooth substances | individual particles | invisible forces |

that _____.

| stay very still | move rapidly |

This is supported by the observation that the balloon _____

| inflates | deflates |

when gas is added. The gas in the balloon could be modeled using _____

| water | rocks | bouncy balls |

in a container.

SEP Developing and Using Models
CCC Scale, Proportion, and Quantity

Concept 1.3: A Model of Matter

1.3 | Share
How can a model show us the behavior of the particles that make up matter?

Activity 14
Record Evidence Like a Scientist

Quick Code: us5063s

Models

Now that you have learned about modeling matter, look again at the video Let's Investigate Models. You first saw this in Wonder.

Let's Investigate Models

Talk Together

How can you describe models now? How is your explanation different from before?

SEP Constructing Explanations and Designing Solutions

Look at the Can You Explain? question. You first read this question at the beginning of the lesson.

> **Can You Explain?**
>
> How can a model show us the behavior of the particles that make up matter?

Now, you will use your new ideas about models to answer a question.

1. **Choose** a question. You can use the Can You Explain? question or one of your own. You can also use one of the questions that you wrote at the beginning of the lesson.

My Question

2. Then, use the ideas that follow to help you answer the question.

Concept 1.3: A Model of Matter | 127

1.3 | Share
How can a model show us the behavior of the particles that make up matter?

A scientific explanation consists of a **claim, evidence** to support the claim, and **reasoning** connecting the evidence to the claim. A scientific explanation of the effect of temperature on the behavior of particles that make up matter would have the following structure.

The **claim** should include how the temperature of an object affects the motion of particles of that object.

Now, **write** your claim:

_____
_____
_____

Next, **list** some of your evidence:

_____
_____
_____
_____
_____

The **reasoning** is why your evidence supports your claim.

Explain your reasoning:

_____

_____

_____

_____

1.3 | Share

How can a model show us the behavior of the particles that make up matter?

Now, **write** your scientific explanation.

 **in Action**

Quick Code: us5064s

 **Activity 15**
Analyze Like a Scientist

Atom Smashers

Read the text, **look** at the image, and **answer** the questions that follow.

Atom Smashers

Particle physicists are scientists that study what atoms are made of. These scientists use a tool called a particle accelerator. Another name for it is "atom smasher"! The atom smasher is a closed, circular tunnel. Inside the atom smasher, streams of charged particles travel around and around with increasing speed. When they are fast enough, they are made to smash into target particles or atoms. The protons, **electrons**, and atoms are smashed into even smaller particles. Machines study the smaller particles that result from the atoms smashing.

As Fast as Light

SEP Obtaining, Evaluating, and Communicating Information

Concept 1.3: A Model of Matter | 131

Atom Smashers *cont'd*

What have particle physicists learned from smashing atoms? They have learned that protons and electrons can be divided into even smaller particles. One of these particles will help them answer the question: Why does matter have mass? Why is this important to know? This will help scientists understand how the universe and life began.

People use high-energy electrons to kill harmful bacteria in foods such as meat, seafood, and spices. Some of the food that astronauts eat is processed this way so astronauts do not get sick in space. Doctors also use electron beams to take pictures of patients' internal parts to find and treat health problems. At this time, scientists are testing to see if proton beams can be used to kill cancer cells. Who would have guessed that tiny, little particles, such as atoms, protons, and electrons, would have such great strength!

Vacuum Sealed

A vacuum in science is different from the vacuum that we use to clean carpets. A vacuum in science is a place where no matter, such as dust, air, or water, can be found. The tube through which the particles pass in an atom smasher is a vacuum. Why might this be important?

_____

_____

_____

Atom Smasher

Electron beams are used to kill harmful bacteria, and proton beams may be used to kill cancer cells. What conclusion can you draw about these particle beams? What property of the beams can support your conclusion?

_____

_____

_____

Design a physical model that shows how atoms are smashed in a particle accelerator and what happens after atoms are smashed. **Draw** and **describe** your model.

Concept 1.3: A Model of Matter | 133

1.3 | Share
How can a model show us the behavior of the particles that make up matter?

Activity 16
Evaluate Like a Scientist

Quick Code: us5065s

Review: A Model of Matter

Think about what you have read and seen in this lesson. **Write** down some core ideas you have learned. **Review** your notes with a partner. Your teacher may also have you take a practice test.

Core Ideas

_____
_____
_____
_____
_____
_____
_____
_____
_____
_____
_____
_____
_____

SEP Obtaining, Evaluating, and Communicating Information

Talk Together

Think about what you saw in Get Started. Use your new ideas about modeling matter to discuss water disappearing from a fishbowl.

Unit Project

Solve Problems Like a Scientist

Unit Project: Decreasing Water Levels

Quick Code: us5066s

Watch the video about a problem involving a fishbowl. What do you think is causing this problem?

Decreasing Water Levels

In this activity, you will investigate the problem presented in the video. Your task is to design an investigation, collect data, graph your data, and analyze the results to explain why the problem is occurring.

SEP	Developing and Using Models
SEP	Constructing Explanations and Designing Solutions
CCC	Cause and Effect

What materials do you need? (per group)

- Beakers, plastic, 250 mL
- Pocket scale
- Thermometer, plastic
- Labels
- Plastic wrap
- Wax pencil

What Will You Do?

1. With your partner, decide on the question you will answer in this investigation. Record your question.

   _____
   _____
   _____

2. With your partner, discuss possible hypotheses that provide an answer to your investigative question. Record one hypothesis that you will test in this investigation.

   _____
   _____
   _____

Unit 1: What Is Matter Made Of?

Unit Project

3. Discuss the procedure that you will follow in your investigation. Write out the steps. Then, have your teacher approve your procedure before you begin.

4. Carry out your investigation, collect data and observations, and record these in the space provided.

5. Use your data to construct 2–4 graphs showing changes in mass and temperature over time in the space provided. Clearly label each part of your graphs, and provide a title and key for each graph.

Think About the Activity

Explain the results from your graphs and how this relates to the particle model of matter.

_____
_____
_____
_____
_____
_____
_____
_____
_____
_____

Grade 5 Resources

- **Bubble Map**
- **Safety in the Science Classroom**
- **Vocabulary Flash Cards**
- **Glossary**
- **Index**

Name _____

Bubble Map

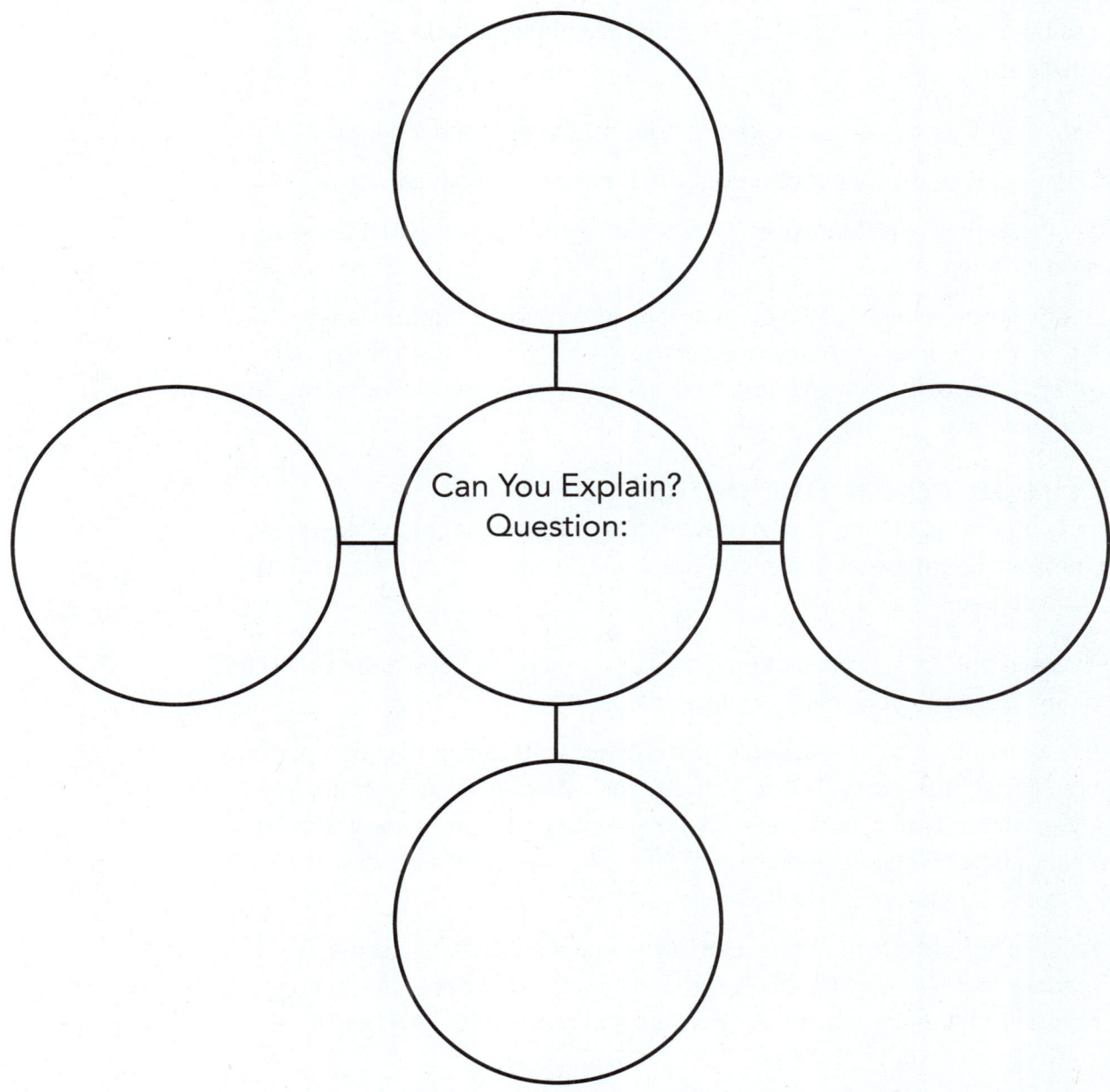

Bubble Map | R3

Safety

Safety in the Science Classroom

Following common safety practices is the first rule of any laboratory or field scientific investigation.

Dress for Safety

One of the most important steps in a safe investigation is dressing appropriately.

- Splash goggles need to be kept on during the entire investigation.
- Use gloves to protect your hands when handling chemicals or organisms.
- Tie back long hair to prevent it from coming in contact with chemicals or a heat source.
- Wear proper clothing and clothing protection. Roll up long sleeves, and if they are available, wear a lab coat or apron over your clothes. Always wear close toed shoes. During field investigations, wear long pants and long sleeves.

Be Prepared for Accidents

Even if you are practicing safe behavior during an investigation, accidents can happen. Learn the emergency equipment location in your classroom and how to use it.

- The eye and face wash station can help if a harmful substance or foreign object gets into your eyes or onto your face.
- Fire blankets and fire extinguishers can be used to smother and put out fires in the laboratory. Talk to your teacher about fire safety in the lab. He or she may not want you to directly handle the fire blanket and fire extinguisher. However, you should still know where these items are in case the teacher asks you to retrieve them.
- Most importantly, when an accident occurs, immediately alert your teacher and classmates. Do not try to keep the accident a secret or respond to it by yourself. Your teacher and classmates can help you.

Practice Safe Behavior

There are many ways to stay safe during a scientific investigation. You should always use safe and appropriate behavior before, during, and after your investigation.

Safety Goggles

- Read the all of the steps of the procedure before beginning your investigation. Make sure you understand all the steps. Ask your teacher for help if you do not understand any part of the procedure.

- Gather all your materials and keep your workstation neat and organized. Label any chemicals you are using.

- During the investigation, be sure to follow the steps of the procedure exactly. Use only directions and materials that have been approved by your teacher.

- Eating and drinking are not allowed during an investigation. If asked to observe the odor of a substance, do so using the correct procedure known as wafting, in which you cup your hand over the container holding the substance and gently wave enough air toward your face to make sense of the smell.

- When performing investigations, stay focused on the steps of the procedure and your behavior during the investigation. During investigations, there are many materials and equipment that can cause injuries.

- Treat animals and plants with respect during an investigation.

- After the investigation is over, appropriately dispose of any chemicals or other materials that you have used. Ask your teacher if you are unsure of how to dispose of anything.

- Make sure that you have returned any extra materials and pieces of equipment to the correct storage space.

- Leave your workstation clean and neat. Wash your hands thoroughly.

Vocabulary Flash Cards

atmosphere

Image: Paul Fuqua

layers of gas that surround a planet

boil

Image: Pixabay

to change state from a liquid to a gas because of added heat

change of state

Image: Discovery Communications, Inc.

the process of changing (by heating or cooling) one form of matter into another form of matter

chemical change

Image: Paul Fuqua

a chemical reaction; a process that changes substances into new substances

combine

to bring together; to mix

electron

a subatomic particle with a negative charge

energy

the ability to do work or cause change; the ability to move an object some distance

gas

a state of matter without any defined volume or shape

heat

Image: Paul Fuqua

the transfer of thermal energy

light

waves of electromagnetic energy; electromagnetic energy that people can see

liquid

Image: Thinkstock / Stockbyte / Getty Images

a state of matter with a defined volume but no defined shape

mass

Image: Andrew Martin/Pixabay

the amount of matter in an object

Vocabulary Flash Cards | R11

material

things that can be used to build or create something

matter

material that has mass and takes up some amount of space

measure

to use a tool to learn more about the volume, length or weight of an object

melt

to change a substance from solid to liquid

Vocabulary Flash Cards | R13

mixture

Image: shell_ghostcage/Pixabay

a combination of substances that can be physically separated from one another

model

a drawing, object, or idea that represents a real event, object, or process

particle

Image: Free-Photos/Pixabay

something that is very tiny

physical change

Image: Christos Giakkas/Pixabay

a change in matter that does not affect its chemical composition

Vocabulary Flash Cards | R15

plasma

a fourth state of matter in which the particles of a gas become highly charged (ionized)

property

a characteristic or quality of a material

scale

a description of the relative size or amount of two or more things; a device used for measuring weight

solid

matter with a fixed volume and shape

state of matter

a particular form that matter can take; the three main states of matter are solid, liquid, and gas

substance

the physical matter of which living or nonliving things are composed

sugar

a chemical compound that organisms use for energy

thermal energy

energy in the form of heat

volume

Image: Discovery Education

the amount of space that matter takes up

warm

Image: Free-Photos/Pixabay

having heat

water

Image: Paul Fuqua

a compound made of hydrogen and oxygen

Vocabulary Flash Cards | R21

Glossary

English ———— A ———— Español

abiotic includes all nonliving things	**abiótico** incluye todos los objetos sin vida
absorb to take in	**absorber** incorporar
abyssal zone the zone of deep ocean below a depth of 4,000 meters	**zona abisal** zona del océano profundo por debajo de los 4,000 metros de profundidad
air the part of the atmosphere closest to Earth; the part of the atmosphere that organisms on Earth use for respiration	**aire** parte de la atmósfera más cercana a la Tierra; la parte de la atmósfera que los organismos que habitan la Tierra utilizan para respirar
aquatic relating to water	**acuático** relativo al agua
Arctic being from an icy climate, such as the North Pole	**ártico** que pertenece a un clima helado, como el Polo Norte

astronaut
a person who travels outside Earth's atmosphere; a person who travels into space (related words: astronomer, astronomy)

astronauta
persona que viaja fuera de la atmósfera de la Tierra; una persona que viaja por el espacio (palabras relacionadas: astrónomo, astronomía)

atmosphere
layers of gas that surround a planet (related word: atmospheric)

atmósfera
capas de gas que rodean un planeta (palabra relacionada: atmosférico)

axis
an imaginary line that an object spins or revolves around

eje
línea real o imaginaria que pasa por el centro de un objeto; el objeto gira alrededor de ella

——— B ———

beaker
a scientific measuring cup used to measure liquids

vaso de precipitados
contenedor cilíndrico de vidrio usado en laboratorio

biodiversity
all of the organisms that live together in an environment

biodiversidad
todos los organismos que viven juntos en un medio ambiente

biosphere
that part of Earth in which life can exist

biosfera
parte de la Tierra donde puede existir la vida

biotic
includes all living things

biótico
incluye a todos los seres vivos

boil
to change state from a liquid to a gas because of added heat

hervir
cambiar el estado de líquido a gas por el aumento de la temperatura

--- C ---

carbon dioxide
a waste product made by cells of the body; a gas in the air made of carbon and oxygen atoms: humans rid themselves of carbon dioxide waste by exhaling, or breathing out

dióxido de carbono
desecho formado por células del cuerpo; gas en el aire que se forma por átomos de carbono y oxígeno: los humanos desechamos dióxido de carbono al exhalar o expulsar aire

celestial sphere
an imaginary sphere surrounding Earth that show the sky and stars overhead

esfera celeste
esfera imaginaria que rodea la Tierra, y que muestra el cielo y las estrellas

Celsius
the metric temperature scale: Water boils at 100 degrees Celsius, and it freezes at 0 degrees Celsius

Celsius
escala de la temperatura métrica: el agua hierve a los 100 grados centígrados y se congela a los 0 grados centígrados

change of state
the process of changing (by heating or cooling) one form of matter into another form of matter with different characteristics

cambio de estado
el proceso de cambiar (mediante frío o calor) una forma de materia a otra forma de materia con diferentes características

chemical change
a process that changes substances into new substances and is different physically from the original substance

cambio químico
proceso que transforma a las sustancias en nuevas sustancias que son diferentes físicamente de la que las originó (palabra relacionada: reacción química)

chemical energy
energy that can be changed into motion and heat, stored in the bonds between atoms

energía química
energía que se puede cambiar a movimiento y calor, y que está almacenada en las cadenas entre átomos

combine
to bring together; to mix (related word: combination)

combinar
unir, mezclar (palabra relacionada: combinación)

conserve
to protect something or prevent the wasteful overuse of a resource

conservar
proteger algo o evitar el uso excesivo e ineficiente de un recurso

constant
continuing without interruption

constante
que continúa sin interrupción

constellation
a particular area of the sky; a group of stars

constelación
área particular del cielo; grupo de estrellas

consumer
an organism that eats other living things to get energy; an organism that does not produce its own food (related word: consume)

consumidor
organismo que come otros seres vivos para obtener energía; organismo que no produce su propio alimento (palabra relacionada: consumir)

crater
a large, circular pit in the surface of a planet or other body in space usually formed when two bodies in space collide

cráter
hoyo grande y circular en la superficie de un planeta u otro cuerpo en el espacio, generalmente formado cuando chocan dos cuerpos en el espacio

cycle
a process that repeats (related word: cyclic)

ciclo
proceso que se repite (palabra relacionada: cíclico)

--- D ---

decomposer
organism that carries out the process of breaking things down into dead or decaying organisms

descomponedor
organismo que lleva a cabo el proceso de descomposición mediante la desintegración de los organismos muertos

E

Earth
the third planet from the sun; the planet on which we live (related words: earthly; earth – meaning soil or dirt)

Tierra
tercer planeta desde el Sol; planeta en el cual vivimos (palabras relacionadas: terrenal; tierra en el sentido de suelo o suciedad)

ecosystem
all the living and nonliving things in an area that interact with one another

ecosistema
todos los seres vivos y objetos sin vida de un área, que se interrelacionan entre sí

electron
a particle with a negative charge

electrón
partícula subatómica con una carga negativa

energy
the ability to do work or cause change; the ability to move an object some distance

energía
habilidad para hacer un trabajo o producir un cambio; habilidad para mover un objeto a cierta distancia

energy pyramid
a model that represents energy transfer within an ecosystem, showing a lot of energy at the bottom and a much smaller amount on top

pirámide energética
un modelo que representa la transferencia de energía dentro de un ecosistema, que muestra mucha energía en la base y una cantidad mucho más pequeña en la cima

energy transfer
the transfer of energy from one organism to another through a food chain or web; or the transfer of energy from one object to another, such as heat energy

transferencia de energía
transmisión de energía desde un organismo a otro a través de una cadena o red de alimentos; o transferencia de energía desde un objeto a otro, como por ejemplo la energía del calor

environment
all the living and nonliving things that surround an organism

medio ambiente
todos los seres vivos y objetos sin vida que rodean a un organismo

estuary
a coastal body of water where freshwater from a river mixes with saltwater from the ocean

estuario
cuerpo costero de agua donde el agua dulce de un río se mezcla con el agua salada del océano

F

food chain
a model that shows how energy passes from one organism to another in an ecosystem

cadena alimentaria
modelo que muestra un conjunto de relaciones de alimentación entre seres vivos en un ecosistema

food web
shows the ways in which many food chains work with one another in an ecosystem

red alimentaria
muestra la manera en que muchas cadenas alimentarias trabajan entre sí en un ecosistema

force
a pull or push that is applied to an object

fuerza
acción de atraer o empujar que se aplica a un objeto

freshwater
water that is not salty, such as that found in streams and lakes

agua dulce
agua que no es salada, como la que se encuentra en arroyos y lagos

friction
a force that stops motion

fricción
fuerza que desacelera o detiene el movimiento

―――― G ――――

galaxy
a group of solar systems, dust, and gas held together by gravity; our solar system is part of the Milky Way galaxy

galaxia
grupo de sistemas solares, polvo y gas unidos por la gravedad; nuestro sistema solar es parte de la galaxia llamada Vía Láctea

gas
a state of matter without any defined volume or shape (related word: gaseous)

gas
estado de la materia sin volumen ni forma definidos (palabra relacionada: gaseoso)

geosphere
Earth's crust, both beneath the oceans and continents, as well as the mantle and inner and outer core

geosfera
corteza terrestre, tanto debajo de los océanos como de los continentes, así como también el manto y los núcleos interior y exterior

glacier
a large sheet of ice or snow that moves slowly over Earth's surface

glaciar
sábana de hielo o nieve que se mueve lentamente sobre la superficie de la Tierra

gravity
the force that pulls an object toward the center of Earth (related word: gravitational)

gravedad
fuerza que empuja a un objeto hacia el centro de la Tierra (palabra relacionada: gravitacional)

H

heat
the transfer of thermal energy

calor
transferencia de energía térmica

horizon
the point at which Earth's surface appears to meet the sky

horizonte
punto en el cual la superficie de la Tierra parece reunirse con el cielo

hydrogen
the most abundant element in the universe, made of one proton and one electron

hidrógeno
el elemento químico más abundante en el universo, hecho de un protón y un electrón

hydrosphere
all of the water on, under, and above Earth

hidrósfera
toda el agua que se encuentra sobre, debajo y en la Tierra

I

imaginary
existing only in your mind or in your imagination

imaginario
que existe sólo en nuestra mente o en nuestra imaginación

interact
to act on one another (related word: interaction)

interactuar
ejercer influencia mutua (palabra relacionada: interacción)

L

light
a form of energy that moves in waves and particles and can be seen

luz
una forma de energía que se mueve en ondas y partículas y que se puede ver

light energy
that form of energy that animals can see directly; visible electromagnetic radiation

energía lumínica
tipo de energía que los animales pueden ver directamente; radiación electromagnética visible

light year
the distance light travels in a vacuum in one year; about 6 trillion miles

año luz
distancia que viaja la luz en el espacio en un año; alrededor de 6 billones de millas

liquid
a state of matter with a defined volume but no defined shape

líquido
estado de la materia que posee un volumen definido, pero no una forma definida

location
a place where something is positioned

ubicación
lugar donde se posiciona algo

---- M ----

magnet
an object with a north and south pole that produces a magnetic field (related terms: magnetism, magnetic)

imán
objeto con un polo norte y un polo sur que produce un campo magnético (palabras relacionadas: magnetismo, magnético)

magnify
to make something appear larger, usually by using one or more lenses

ampliar
hacer que algo parezca más grande, generalmente usando una o más lentes

mass
the amount of matter in an object

masa
cantidad de materia en un objeto

material
things that can be used to build or create something

material
cosas que se pueden usar para construir o crear algo

matter
material that has mass and takes up some amount of space

materia
material que tiene masa y ocupa cierta cantidad de espacio

measure
to use a tool to learn more about the volume, length, or weight of an object (related word: measurement)

medir
usar una herramienta para saber más sobre el volumen, la longitud, o el peso de un objeto (palabra relacionada: medición)

mechanical energy
energy that an object has because of its motion or its position

energía mecánica
energía que tiene un objeto debido a su movimiento o posición

melt
to change a substance from solid to liquid

derretirse
cambio de sustancia de estado sólido a líquido

mineral
a natural, solid substance found in rocks; each mineral has a specific chemical makeup

mineral
sustancia natural y sólida que se encuentra en las rocas; cada mineral tiene una composición química específica

mixture
a combination of substances that can be physically separated from one another

mezcla
combinación de sustancias que pueden separarse físicamente unas de otras

model
a drawing, object, or idea that represents a real event, object, or process

modelo
dibujo, objeto o idea que representa un evento, objeto, o proceso real

moon
a body in outer space that orbits a planet; a natural satellite

luna
cuerpo en el espacio exterior que gira alrededor de un planeta; satélite natural

motion
a change in the position of an object compared to another object (related terms: move, movement)

movimiento
cambio en la posición de un objeto en comparación con otro objeto (palabras relacionadas: mover, desplazamiento)

N

natural resources
resources that are obtained from Earth

recursos naturales
recursos obtenidos de la Tierra

nebula
an interstellar cloud made up of hydrogen gas, plasma, helium gas, and dust

nebulosa
una nube interestelar constituida por gases (hidrógeno y helio), plasma, y polvo

nonrenewable
once it is used, it cannot be made or reused again

no renovable
una vez usado, no puede rehacerse o reutilizarse

nonrenewable resource
a natural resource of which a finite amount exists, or one that cannot be replaced with currently available technologies

recurso no renovable
recurso natural del cual existe una cantidad finita, o uno que no puede remplazarse con las tecnologías actualmente disponibles

nuclear energy
the energy released when the nucleus of an atom is split apart (during fission) or combined with another nucleus (during fusion)

energía nuclear
energía liberada cuando el núcleo de un átomo se divide (durante la fisión) o combina con otro átomo (durante la fusión)

nucleus
the center of an atom containing protons and neutrons (related terms: nuclei, nuclear); a region in a cell that is surrounded by a membrane and contains genetic material

núcleo
centro de un átomo que contiene protones y neutrones (palabras relacionadas: núcleos, nuclear); región de una célula que está rodeada por una membrana y contiene material genético

nutrients
important particles found in food that a living thing needs to survive

nutrientes
importantes partículas que se encuentran en los alimentos y que un ser vivo necesita para sobrevivir

O

ocean
a large body of saltwater that covers most of Earth

océano
gran cuerpo de agua salada que cubre la mayor parte de la Tierra

optical
having to do with the eye or lenses

óptico
relacionado con los ojos o con las lentes

orbit
the circular path of an object as it revolves around another object

órbita
trayectoria circular de un objeto que se forma a medida que gira alrededor de otro objeto

organism
any individual living thing

organismo
todo ser vivo individual

oxygen
a gas found in air and water that living things need to breathe

oxígeno
gas que se encuentra en el aire y en el agua, y que los organismos vivos respiran

P

particle
something that is very tiny

partícula
algo que es muy pequeño

perpendicular
a downward direction, making a right angle

perpendicular
con dirección descendente, formando un ángulo recto

phenomenon
a scientific wonder or happening

fenómeno
una maravilla o suceso científico

physical change
a change in matter that does not change the makeup of a substance

cambio físico
alteración en la materia que no afecta su composición química

plankton
small organisms that drift through bodies of water; include animals, plants, and bacteria

plancton
organismos pequeños que se mueven a través de cuerpos de agua; incluyen animales, plantas, y bacterias

plant
an organism that is made up of many cells, makes its own food through photosynthesis, and cannot move; a member of kingdom Plantae

planta
organismo formado por muchas células que fabrica su propio alimento a través de la fotosíntesis, y no se puede mover; miembro del reino vegetal

plasma
a fourth state of matter in which the particles of a gas become highly charged (ionized)

plasma
cuarto estado de la materia en el que las partículas de un gas llegan a estar altamente cargadas (ionizadas)

pollute
to put harmful materials into the air, water, or soil (related words: pollution, pollutant)

contaminar
poner materiales perjudiciales en el aire, agua, o suelo (palabras relacionadas: contaminación, contaminante)

precipitation
water that is released from clouds in the sky; includes rain, snow, sleet, hail, and freezing rain

precipitación
agua liberada de las nubes en el cielo; incluye la lluvia, la nieve, la aguanieve, el granizo, y la lluvia congelada

prey
an animal that is hunted and eaten by another animal

presa
animal que es cazado y comido por otro

produce
to make or create something

producir
hacer o crear algo

producer
an organism that makes its own food; an organism that does not consume other plants or animals

productor
organismo que fabrica su propio alimento; organismo que no consume otras plantas u otros animales

property
a characteristic or quality of a material

propiedad
una característica o calidad de un material

—— R ——

radiant energy
energy that does not need matter to travel; light

energía radiante
energía que no necesita de la materia para viajar; luz

radiation
electromagnetic energy (related word: radiate)

radiación
energía electromagnética (palabra relacionada: irradiar)

recycle
to create new materials from used products

reciclar
crear nuevos materiales a partir de productos usados

refuse
garbage

desecho
basura

rely
to be dependent upon

depender
estar sujeto a, estar subordinado a

renewable resource
a natural resource that can be replaced

recurso renovable
recurso natural que puede reemplazarse

resource
a naturally occurring material in or on Earth's crust or atmosphere of potential use to humans

recurso
material que se origina de forma natural en o sobre la corteza o la atmósfera de la Tierra, que es de uso potencial para los seres humanos

revolution
the orbiting of an object around another object

revolución
movimiento por el cual un objeto gira alrededor de otro objeto describiendo una órbita completa

river
big body of water flowing through land on either side

río
gran cuerpo de agua que fluye con tierra de ambos lados

rotate
turning around on an axis; spinning (related word: rotation)

rotar
girar sobre un eje; dar vueltas (palabra relacionada: rotación)

rotation
the spinning of a celestial body, such as a planet, around an axis

rotación
giro de un cuerpo celeste, como un planeta, alrededor de un eje

---------- **S** ----------

salt
a mineral found in the ocean and other parts of Earth that can be used for preserving things and seasoning food

sal
un mineral que se encuentra en el océano y otras partes de la Tierra, que se puede usar para conservar cosas y condimentar alimentos

saltwater
contains salt and other minerals that make the water unsuitable for drinking

agua salada
contiene sal y otros minerales que hacen que el agua no sea apta para beber

scale
a device used for measuring weight

escala/báscula
dispositivo empleado para medir el peso

shelter
a location or structure that gives protection from danger or bad weather

refugio
lugar o estructura que brinda protección contra los peligros o las malas condiciones meteorológicas

solar system
a system of objects that revolve around a star

sistema solar
conjunto de objetos que giran alrededor de una estrella

solid
matter with a fixed volume and shape

sólido
materia con un volumen y una forma determinada

sound
anything you can hear that travels by making vibrations in air, water, and solids

sonido
vibración que viaja a través de un material, como el aire o el agua

star
a massive ball of gas in outer space that gives off heat, light, and other forms of radiation

estrella
bola masiva de gas en el espacio exterior que emite calor, luz, y otras formas de radiación

state of matter
a particular form that matter can take: the three main states of matter are solid, liquid, and gas.

estado de la materia
forma particular que puede tener la materia: los tres estados principales de la materia son sólido, líquido, y gaseoso

stem
the part of a plant that grows away from the roots; supports leaves and flowers

tallo
parte de la planta que crece a partir de la raíz; contiene las hojas y la flores

stomata
pores on the surface of a plant that allow gases to move into and out of the plant (related word: stoma)

estomas
poros en la superficie de una planta que permiten a los gases moverse hacia adentro y hacia afuera de la planta (palabra relacionada: estoma)

stream
a small body of flowing water

arroyo
pequeño cuerpo de agua que fluye

substance
the physical matter of which living or nonliving things are made

sustancia
materia física de la cual están compuestos los seres vivos y objetos sin vida

sugar
a substance that is sweet, mainly used in food and drinks

azúcar
sustancia dulce, utilizada principalmente en alimentos y bebidas

sun
any star around which planets revolve

sol
toda estrella alrededor de la cual giran los planetas

surface
the top or outermost part of something

superficie
parte superior o externa de un objeto

survive
to continue living or existing: an organism survives until it dies; a species survives until it becomes extinct (related word: survival)

sobrevivir
continuar viviendo o existiendo: un organismo sobrevive hasta que muere; una especie sobrevive hasta que se extingue (palabra relacionada: supervivencia)

sustainable
able to be used over and over again without hurting the overall supply

sostenible
que se puede utilizar una y otra vez sin afectar el suministro total

system
a group of related objects that work together to perform a function

sistema
grupo de objetos relacionados que funcionan juntos para realizar una función

T

telescope
an instrument used to observe objects that are far away

telescopio
instrumento usado para observar objetos que se encuentran alejados

terrarium
a human-made ecosystem used for organisms to grow, usually in a clear container to allow for observation

terrario
ecosistema creado por el hombre para que crezcan organismos, generalmente en un recipiente transparente para permitir la observación

thermal energy
energy in the form of heat

energía térmica
energía en forma de calor

U

universe
everything that exists in, on, and around Earth

universo
todo lo que existe en, sobre, o alrededor de la Tierra

V

volcano
an opening in Earth's surface through which magma and gases or only gases erupt (related word: volcanic)

volcán
abertura en la superficie de la Tierra a través de la cual el magma y los gases o sólo los gases hacen erupción (palabra relacionada: volcánico)

volume
the amount of space that matter takes up

volumen
la cantidad de espacio que ocupa la materia

W

warm
having heat

cálido
que tiene calor

water
a compound made of hydrogen and oxygen; can be in either a liquid, ice, or vapor form and has no taste or smell

agua
compuesto hecho de hidrógeno y oxígeno, puede presentarse en forma líquida, en forma de hielo o vapor y no tiene sabor ni olor

watershed
a region in which all precipitation and surface water collects and drains into the same river

cuenca
región en la que se recoge toda la precipitación y agua superficial, que drena hacia el mismo río

weight
the force of gravity on an object

peso
fuerza de gravedad que se ejerce sobre un objeto

Index

A

Analyze Like a Scientist 17–18, 28–29, 31–32, 43–45, 63–65, 67–69, 80–81, 90–92, 97–100, 111, 113–114, 115–116, 122–123, 131–133
Ask Questions Like a Scientist 10–11, 52–53, 106–107
Atmosphere 67
Atom 131–133

B

Boil 90

C

Can You Explain? 8, 41, 50, 95, 104, 127
Change of state 64, 90–91
Chemical
 properties of 28–29, 31
 mixtures 67–68
 change 80–81, 90–91
 state 63–64
Combine 68, 81
Conduction 122–123

E

Electricity 32
Electron 113, 131–132
Energy 17, 63–64, 116, 122–123
Evaluate Like a Scientist 12–13, 20–21, 30, 33, 46–47, 56–57, 78–79, 88–89, 93, 101, 109, 112, 117–118, 124, 134–135
Evaporation 68, 123

F

Filter 68
Freezing 80, 97, 123

G

Gas 17, 67, 80, 90, 97, 111, 114

H

Hands-On Activities 24–27, 34–39, 58–62, 70–77, 84–87
Heat 81, 116

I

Investigate Like a Scientist 24–27, 34–39, 58–62, 70–77, 84–87

L

Light 116
Liquid 17, 64, 72, 90–91, 97, 111

M

Mass 17, 19, 29, 90–91, 93, 111, 116, 132
Material 28, 43
Matter 12, 14, 17, 22, 28–31, 33, 43–44, 46, 56–58, 63, 67, 80, 82, 94, 97, 109–112, 116, 119, 122, 132, 134
Measure 18, 24, 43–45
Melt 52, 64, 94
Mixture 66–68, 70, 78
Model 106, 108–109, 112, 117, 119, 122, 126, 134
Motion 115–117, 122–123

O

Observe Like a Scientist 14–16, 19, 22–23, 54–55, 66, 82–83, 108, 110

P

Particles 17, 29, 63, 111–116, 122–123, 131
Physical 31, 32, 64, 80, 84, 88, 91
Property 14, 24, 28, 43–44, 68, 78

R

Record Evidence Like a Scientist 40–42, 94–96, 126–130

S

Scale 18
Solid 17, 54, 64, 71–72, 80, 97, 111
Solve Problems Like a Scientist 4–5 136–139
State
    boiling 80, 90, 97
    change of 64, 90–91
    condensation 123
    evaporation 68, 123
    freezing 80, 97, 123
    gas 17, 67, 80, 90, 97, 111, 114, 124
    liquid 17, 54, 64, 72, 90–91, 97, 111
    melting 52, 64, 94
    of matter 17, 19, 22, 57, 63–64, 97–98, 111
    solid 17, 54, 64, 71–72, 80, 97, 111
STEM in Action 43–45, 97–100, 131–133
Substance 29, 63, 80–81
Sugar 68, 80

T

Temperature
    heat 81, 116
    and thermal energy 116, 123
    properties 18, 29, 34, 40, 63, 64
    warm 116, 123
Thermal energy 116
Think Like a Scientist 119–121

U

Unit Project 4–5 136–139

V

Vacuum 133
Volume 29, 43, 44

W

Water 64, 80, 90, 91, 97, 123